Putting a New Spin on Groups

The Science of Chaos

Putting a New Spin on Groups

The Science of Chaos

Bud A. McClure
University of Minnesota, Duluth

LEA LAWRENCE ERLBAUM ASSOCIATES, PUBLISHERS
1998 Mahwah, New Jersey   London

Lawrence Erlbaum Associates, Inc., Publishers
10 Industrial Avenue
Mahwah, New Jersey 07430

Cover design by Kathryn Houghtaling Lacey

Library of Congress Cataloging-in-Publication Data

McClure, Bud A.
   Putting a new spin on groups : the science of chaos / Bud A.
McClure.
      p.   cm.
   Includes bibliographical references and index.
   ISBN 0-8058-2904-0 (cloth : alk. paper). — ISBN 0-8058-2905-9
(pbk. : alk. paper)
   1. Social groups. 2. Social interaction. 3. Leadership.
I. Title.
HM131.M3767 1998
305—dc21   98-13600
                                                    CIP

Books published by Lawrence Erlbaum Associates are printed on acid-free paper,
and their bindings are chosen for strength and durability.

Printed in the United States of America
10 9 8 7 6 5 4 3 2 1

*For Deborah, Honey, Eli, and Keegan, who brought order into my life,
and for Buddy, who reintroduced chaos*

Contents

Preface

Most groups never reach their full potential. In many cases, groups never progress beyond the initial stages of development. There is a combination of factors that accounts for this inadequate development, but foremost is the group leader's lack of understanding of group dynamics. In part, group leaders are insufficiently trained and have too few supervised group experiences. In addition, many textbooks fail to satisfactorily characterize the stages of group development. Furthermore, many group phenomena go unexplained.

When I began teaching group dynamics, I used standard textbooks. However, as my group experiences increased, I began to notice certain group phenomena that occurred, which were not described or were unreported in these books. So I began to write about my observations and to refine them as I observed more group activity. Until recently, these articles were loosely related and hung together under the broad rubric of group dynamics. This was before I made two discoveries. The first was the work of Arthur Young and his theory of evolution, which provided me with a framework for many of my ideas about group development. The second occurred as I became more interested in the factors that contributed to systemic change in groups. In my research, I encountered chaos theory. Then, much of what I experienced and thought about group evolution fit together under this emerging paradigm. This book is the result of these discoveries.

Young's model combined with ideas drawn from chaos theory, in addition to forming the foundation, provide the threads that connect the various parts of this book. Each chapter details a little-discussed group phenomenon that is woven into the whole fabric by these two threads.

The pattern that emerges in this quilt is brightly colored and illuminates many of the experiences that leaders encounter in working with groups.

This book is intended to challenge orthodox and static ideas about small-group dynamics. A primary goal of this volume is to offer an alternative model of group development that addresses three factors. First, the model integrates old ideas from previous models of group development and new concepts from chaos theory with the work of Arthur Young. Second, this book emphasizes the importance of conflict in group development and recognizes that group growth, although progressive, is neither linear or unidimensional. Third, particular attention is focused on how groups change, evolve, and mature. Of equal importance is the goal of highlighting certain group phenomena that were given only cursory attention in many group textbooks. These areas include women in authority, group metaphors, regressive groups, and the transpersonal potential of small groups.

This volume is divided approximately in half between a comprehensive examination of group development and the stages that characterize development and subjects that have not normally been the focus of small-group textbooks. The first five chapters of this book include an overview of chaos theory, an extensive presentation of a group development model, detailed description of each stage of that model, and the factors that advance and hinder change. Leadership attributes necessary for effective group facilitation are described in detail. The second half of this volume covers material rarely found in group books. This portion provides an explanation for many phenomena that group leaders encounter, but rarely understand. To facilitate the writing process, I have alternated the use of masculine and feminine pronouns from one chapter to another. The pronouns are interchangeable.

Chapter 1 provides background to the origins of chaos theory. The study of dynamics is examined and the terminology of chaos theory introduced.

Chapter 2 introduces several innovative studies to illustrate how chaos theory is currently being applied in the social sciences. Two important tenets of the theory—self-organization and behavior in systems far from equilibrium—are detailed and form the basis for subsequent discussion of how groups change at all levels of organization.

Chapter 3 reviews group stage theories and introduces Arthur Young's theory of evolution. Young's seven stages of evolution form an arc that is divided into two phases—a descent and an ascent. For Young, the *arc* represents the process evolution undergoes as it moves progressively from complete freedom of movement, through a series of stages that constrain it into permanence, at which point it is propelled back upward to complete freedom. Young's arc forms the basis for an alternative model of group development that contains seven stages. The stages are arrayed on both sides of the arc and joined at the vertex by the critical Conflict–Confrontation Stage. Group development is first depicted as moving through several stages of constraint in which individual identities are temporarily relinquished for the sake of the forming group. After the group navigates the crucial conflict period, freedom is regained in the final stages where individual identities reemerge.

Chapter 4 integrates many of Young's ideas with those from chaos theory and applies them to change and transformation in groups. Group development is characterized by periods of relative calm punctuated by intervals of chaotic activity. This order–chaos–order cycle is essential for growth and reorganization because without undergoing periodic upheaval, groups cannot evolve. Understanding how groups undergo this metamorphosis is essential for effective group leadership because attempts to control and limit it lead to regressive and potentially destructive solutions. Constructs such as phase locking and constructive and destructive interference are used to explain the process of change in groups.

Chapter 5 provides a broad overview of effective group leadership characteristics. Many of the ideas presented in the first several chapters are translated to practical leadership strategies. Leader interventions are divided into two categories—containment or perturbation. Pattern recognition, sensitivity to nuance, and amplification are several leadership skills that fall in one of these two categories. (Each one is fully explained in this chapter.) The concept of high leverage points is introduced and developed as an intervention opportunity at which point group leaders can maximize their influence on the direction of group development. Gender issues relative to group leadership are explored and several factors relative to women in authority are uncovered.

Chapter 6 emphasizes the leadership skills that are stage specific. Particular emphasis is placed on the leader's role in Stage Four, Conflict–Confrontation. Scapegoating is introduced in this chapter, and strategies for leadership intervention to thwart it are discussed. Case examples are used to illuminate discussion.

Chapter 7 includes a transcript of two group sessions that center on a confrontation with a group leader. Detailed explanations of the leader and member interactions are provided from both a psychological perspective and from the perspective of a dynamic system undergoing rapid and discontinuous change. In this analysis, chaos theory and its application to groups is detailed with precision through each member interaction.

Chapter 8 introduces the subtle and symbolic level of group interaction. Group metaphors are defined as analogies that permit group members to remove affect from an emotionally charged situation, substituting a nonthreatening external subject for a threatening internal one. The group metaphor is also examined from the perspective of chaos theory and is equated with a strange attractor. In other words, the group metaphor is the resulting pattern that emerges as the group attempts to resolve over-whelming anxiety. Methods for using group metaphors are explained, and many case examples are provided.

Chapter 9 examines the dark, denied, and unacknowledged behavior of groups and organizations. These groups are labeled regressive and re-main stuck in the forming stages of development. In the language of chaos, regressive groups form a limit-cycle attractor, unable to evolve or develop. These groups remain dependent on the leader for direction, repress anger and dissent, and create out-groups onto which they project their shadows. This chapter examines the development of regressive group characteristics and provides suggestions for transforming these rigid groups into more productive organizations. Leader behaviors that can liberate regressive groups are enumerated. Two extensive case examples are provided.

Chapter 10 examines very high levels of group development and cohe-sion that lead to spiritual and transpersonal growth. Utilizing case studies, the chapter focuses on how group leaders can recognize transpersonal issues and promote spiritual healing.

Chapter 11 provides an overview of various teaching paradigms utilized in most group courses. My group experiences, both as a graduate student and as an instructor, are included to highlight the discussion. The chapter includes discussion of the ethical concerns raised by requiring students to participate in a group experience as part of a course requirement.

ACKNOWLEDGMENTS

A book is never a singular project and I'm grateful to those who have supported this one: my colleagues in the Psychology Department at the University of Minnesota, Duluth, who supported a leave of absence so that I could finish this book; Len Borsari and John Cox of Springfield College who, with expert mentoring, introduced me to groups; Martha Eberhart, my favorite librarian, who enthusiastically helped with my research; Sister Donna Marie whose uncanny insight into group processes helped clarify material in chapter 7; my summer group of Leon Molstad, Bob Peterson, Harry Newby, Sr., Jim Crowley, and Brad Putnam who added balance to my life during the writing process; Kristin Ihlenfeld who brought music to my life, especially the work of Leonard Cohen, just when I needed it; and Barbara Wieghaus, senior production editor at Lawrence Erlbaum Associates, who worked diligently on this project. I am also grateful for all of my former and current students who challenged and field tested many of the ideas in this book. In particular, Dr. Geri Miller and Dr. Phyllis Bengston provided valuable guidance in leading and supervising groups. To all of you, thanks!

—Bud A. McClure

Chapter 1

Chaos

Defining chaos theory and the concepts that comprise it is much "like trying to grasp Jell-O®."[1] Although Chamberlain's analogy to Jell-O was intended to depict the difficulty in defining the strange attractor, her creative description equally captures the cognitive experience of one's first, second, or even third encounter with chaos theory:

> It's easy to see that there is some substance there, that the substance has some specific form, and that it appears solid. When one tries to actually pick some up, however, it quickly becomes a challenge to manage and is transformed into a very different substance than it appeared while sitting on the plate.[2]

Getting a firm grasp on a precise definition of chaos is even difficult for those hip-deep in Jell-O because descriptions of chaos theory vary slightly; in chemistry, it is used to describe dissipative structures; in physics, it is applied to dynamic systems; and in mathematics, it describes fractal geometry.[3] Each discipline speaks a different language, resulting in slightly different meanings for the concepts of chaos. Several years ago someone estimated that there were approximately 31 different definitions of chaos theory. In spite of the difficulty in deciphering the language, it is worth the effort because even a rudimentary understanding of chaos confirms many of the intuitive feelings you might have about how groups work.

Although some of the information presented in this chapter may seem oblique to the intended purpose of clarifying chaos theory (for use with groups), it is important to include in order to give a sense of how the application of chaos theory is relevant across all disciplines and may, indeed, turn out to be the third stage in the evolution of science, as Loye and Eisler asserted.

They depicted the evolution of chaos theory with a three-stage model of science. Stage 1 represented the physics of Aristotle that concentrated on steady or equilibrium states. Stage 2 began with the identification and study of periodic fluctuation, albeit those system states that fluctuate yet remain near equilibrium. Today, in Stage 3, extreme instability and chaos, or states far from equilibrium, are under investigation.[4] This new science of chaos is causing a great deal of excitement in the natural sciences, because unlike Newton's ordered universe—where everything works with clocklike precision—chaos theory embraces the irregularities and the constantly changing side of life. This is a world we feel to be more real than the orderliness purported in the Newtonian universe because it closely matches our own daily encounters with chance, randomness, and uncertainty. Yet, what chaos theory uncovered is that the randomness is not without order, because when scientists map life's irregularities, patterns emerge.

I am reminded here of Bateson's (1989) book, *Composing a Life*, wherein she chronicled the lives of five women. Noticeable in each woman's life, as in our own, were the patterns that were revealed from the perspective of hindsight. What was experienced over the life span of each woman as a series of discontinuous threads of personal and career choices was revealed to be a beautiful woven tapestry. Life's upsets, apparent detours, and difficult decisions all combined to yield an emergent pattern: "The whole fabric of a person's life interacts with his or her total environment, and something new emerges that wasn't predictable from previous behavior."[5]

Chaos theory describes living systems in process and is part of a larger discipline known as nonlinear dynamics. Although scientists investigate chaos from several different perspectives—bifurcation theory, catastrophe theory, the science of complexity—they all study nonequilibrium in dynamic systems.

There are two branches in nonequilibrium theory. The first is the narrowly focused mathematical study of dynamics. It is highly technical and limited to a few specialists.[6] The second and broader area emerges from an effort to find application for these mathematical models in the natural sciences.[7] The interpretation of chaos theory used in this volume comes from this second branch.

A major focus of this volume is small-group transformation. Specifically, what are the factors and conditions that influence group change and development? The answer, I believe, can be found in studying two interconnected ideas generated from the study of chaotic systems—self-organization and nonequilibrium theory.[8] Ilya Prigogine's novel ideas about dissipative structures form the basis of this exploration. His insight into the relationship between order and chaos won him the Nobel Prize in Chemistry in 1984. His description of change in systems far from equilibrium is one of the cornerstones of self-organizing theory used in this volume.

This chapter provides background to the origins of chaos theory. It examines the study of dynamics and introduces the terminology of chaos. Several innovative studies are included to illustrate how chaos theory is currently being utilized in the social sciences.

CHAOS UNCOVERED

Lorenz, studying weather systems on his computer, provided the original mathematical model for explaining unpredictability in the world of nonlinearity.[9] Serendipitously, while trying to predict long-range weather patterns, Lorenz uncovered the elegance of chaos. Perhaps what Lorenz really saw on his computer screen that morning in Massachusetts was the inherent aesthetic of nature: the simple elegance of complexity and nonlinearity that makes possible the vast emergent diversity that we call *evolution*. The importance of his discovery lay buried for almost a decade; then, with the advent of supercomputers, scientists began to model and map turbulent behavior in air and water. Chaos science is now being heralded as a universal language of nature.[10]

In the 17th century, Newton theorized that, given sufficient information about a dynamic system, its behavior could be completely determined by the laws of motion. In part, this is true. For example, solar and lunar eclipses can be predicted decades in advance. This is *deterministic physics.* Here, Newton's calculations are useful. However, his equations could not explain, nor could they account for, random or chaotic behavior.[11] For many years, science simply discounted or overlooked messy and unpredictable behaviors.

Waterfalls, crashing waves, dripping faucets, cloud patterns, heart arrhythmias, fluctuations of the stock market, and predator–prey relationships exhibit a randomness that makes their behavior unpredictable.[12] This random behavior was largely ignored until the emergence of chaos science. Scientists studying chaos proposed that although certain behaviors in natural systems are not predictable, there is a pattern to their randomness or irregularity that emerges over time. To fully comprehend chaos theory, it is necessary to have a rudimentary understanding of dynamics.[13] As mentioned earlier, the mathematics of dynamics forms one of the cornerstones of chaos and self-organizational theories.

DYNAMICS

Dynamics, or the way in which systems change, are characterized as either linear or nonlinear. Linear dynamics, like linear equations, are additive. Therefore, the way a linear system changes can be measured by simply adding the solutions of two or more equations to form a final solution. Linear equations form the basis for the statistics used in the social sciences.[14] Multiple regression, analysis of variance, and log-linear regressions all use additive equations to describe the relationships among variables. Most research is based on linear models because in the real world, linear systems just do not exist. So, opening the door to nonlinear dynamics reveals a worldview biased by linear assumptions.[15] In the nonlinear world, just about anything is possible. Nonlinear systems are very versatile. Therefore, linear equations are limited in their utility for describing systemic changes in nonlinear systems. Nonlinear changes are discontinuous and cause sudden jumps in behavior, making them impossible to quantify with additive equations.

Here is an example that demonstrates both linear and nonlinear change: Begin alternately tapping your index fingers on the table. Within certain limits, you can linearly change or adjust the speed of that tapping to maintain an alternative rhythm. Try changing speeds. Now tap as fast as you can. Notice what happens. Once you exceed a critical speed, you suddenly experience a nonlinear jump and find your fingers tapping in phase.[16] The versatility of nonlinear systems means a single system may show different, even opposing, forms of behavior. "The classic examples here are horses' gaits: walking, trotting, galloping, and running. Each gait represents a completely different organization of leg motion."[17] The transition between gaits, like the leap in finger tapping, occurs in a sudden reorganization.

In nonlinear systems, cause and effect are not functionally related. "A change in one variable may cause proportional changes in other bodies up to a point, at which time the second variable's reactions become nonproportional . . . the straw that broke the camel's back is an apt analogy."[18] Any system in which input is not proportional to output is nonlinear. If you have a headache and take one aspirin, you will get relief; two, more relief; four, perhaps even more, but take 50 aspirins and you will not get as much relief as you did from only one. Thus, a headache is a nonlinear system. "It (nonlinearity) is everything whose graph is not a straight line— and that is essentially everything."[19] In these cases, nonlinear applications are necessary, but they are nonadditive and very difficult to solve.[20] Measuring nonlinear systems lies in discovering their pattern or patterns.

One method of measuring nonlinear systems is Poincaré's mathematical mapping. Marion provides an excellent synthesis of this math.[21] Other methods use what is called iteration to find a system's pattern. After each repetition, the solution is fed back into the original equation. Over time, as the system settles down, it converges into one of four patterns. These patterns are called attractors. Attractors can refer to a discrete point, a simple oscillating cycle, a quasiperiodic or limit cycle, and a chaotic cycle.[22]

ATTRACTORS

The term *attractor* is a bit of a misnomer. The word suggests a magnet pulling objects, or in this case orbits, toward it to organize the system. However, "the attractor is the product of the organization. Mathematically and physically, it is the point that completely describes the state of the system at that particular moment."[23] If the movement of a pendulum was plotted on a graph, a spiral would be created as the pendulum moved back and forth slowly impeded by friction. When the pendulum comes to rest, it is depicted on the graph by a point. The point is the attractor around which the moving pendulum orbited. As systems become more complex, attractors appear as images or forms.

There are differences between linear and nonlinear system attractors.[24] The orbits of two-dimensional linear equations, plotted over time, appear as circles, ellipses, and parabolas—regular and periodic (three-dimensionally, they appear as smooth spheres). However, nonlinear differential equations, when plotted on a computer, produce strange shapes and images (attractors) that move as the system changes. The images depict the system in process. Unlike linear equations, these system orbits, despite coming very close to one another, do not overlap.

Essentially, an attractor refers to the trajectory to which motion gravitates. It has three components. First, an attractor is stable, and any trajectories in its vicinity advance toward it. Pendulum motion exhibits two steady attractors: back-and-forth motion (a periodic attractor) and no motion (a point attractor).[25] Second, an attractor is finite; its behavior is confined within its boundaries. Third, according to its classical definition, an attractor is periodic or quasiperiodic, meaning that its behavior is roughly repetitive, in which subsequent motion nearly returns to its original trajectory.[26] The proof that the attractor does indeed exhibit these characteristics—finite, stable, yet nonrepetitive—is found in Poincaré's mathematical mapping.

In summary, the fixed point attractor (Fig. 1.1) represents the system at rest. After chemicals have stopped their reacting, they settle into a steady state, reflected by a point on a graph. The periodic, quasiperiodic, or limit cycle attractor (Fig. 1.2) is reflected by the motions of a pendulum or metronome and the regular beating of the heart. Its cycle is finite; like the fixed-point attractor, it is regular and predictable. In fact, all future

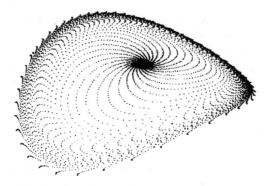

FIG. 1.1. A fixed point attractor that represents a "systems movement toward rest." From *Strange Attractors: Creating Patterns in Chaos* by J. C. Sprott (1993). Copyright © 1993 by J. C. Sprott. Reprinted by permission of Henry Holt and Co., Inc.

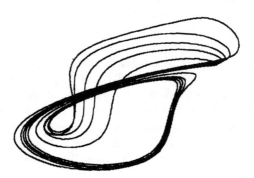

FIG. 1.2. A cyclic or limit cycle attractor that represents a stable system whose dynamics are cyclic. From *Strange Attractors: Creating Patterns in Chaos* by J. C. Sprott (1993). Copyright © 1993 by J. C. Sprott. Reprinted by permission of Henry Holt and Co., Inc.

states of both fixed-point and periodic attractors can be mapped if you know their initial conditions.[27]

For many years, the periodic and fixed-point attractors were thought to form the set. Then, Lorenz discovered another named the *strange attractor*[28] (Fig. 1.3), meaning idiosyncratic, because no two patterns are ever alike, and although it has a geometric shape and is finite and stable, it is neither periodic or quasiperiodic. The strange attractor maps the changing behavior of a dynamic system. By charting the points of a system that reflect its position at different times, the system can be chronicled as it moves and changes. The shape or map that emerges from these plotted points reveals the chaos pattern of the strange attractor. Although the behavior it rep-

FIG. 1.3. A strange attractor that depicts a system with more complex dynamics. While the behavior reflected in this attractor never repeats and exhibits greater variety than either a fixed point or cyclic attractor, it still produces a recognizable pattern. From *Strange Attractors: Creating Patterns in Chaos* by J. C. Sprott (1993). Copyright © 1993 by J. C. Sprott. Reprinted by permission of Henry Holt and Co., Inc.

resents never repeats itself, it somehow manages to stay within specific boundaries. Thus, in the midst of apparent chaos and disorder, these strange attractors symbolically represent some underlying order or pattern. Apparently, certain systems in nature have a kind of internal clock or structure that allows them to maintain order amid chaos. That structure is manifested in the images and patterns of strange attractors. Later in this volume, group metaphors are discussed as a form of strange attractor.

These attractor patterns (Fig. 1.4) are quite intricate, elegant, and beautiful, having been described as "baroque spirals, elaborate filigrees, intricate webs spun by non-Euclidean spiders, shapes like amusement-park rides as depicted by Marcel Duchamp."[29] A popular example of strange attractor

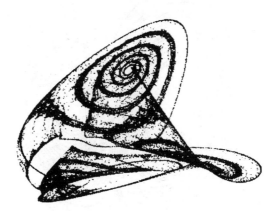

FIG. 1.4. Strange attractor. From *Strange Attractors: Creating Patterns in Chaos* by J. C. Sprott (1993). Copyright © 1993 by J. C. Sprott. Reprinted by permission of Henry Holt and Co., Inc.

motion is boiling water. As water is heated, periodic wiggles appear that constitute a limit cycle. However, as more heat is added, the pattern doubles, creating wiggles on wiggles until a critical point is reached where the water shifts into strange attractor motion.[30] Out of what, at first, appears to be random or chaotic activity, emerge quirky shapes and patterns that follow certain rules and numerical constraints. More amazing, however, is the fact that the laws that explain the chaotic behavior of boiling water also apply to all nonlinear systems. Thus, nonlinear equations, as we see shortly, can model brain-wave behavior, chemical reactions, even social systems. In fact, recent attempts have been made to model the behavior of small group dynamics.[31]

CHAOS TERMINOLOGY

Before we examine applications of chaos theory in the social sciences, let us define several important concepts that have emerged from nonlinear dynamics. A cornerstone of chaos theory, discovered by Lorenz, is known as the Butterfly Effect.[32] It is aptly named because of the pattern it creates when graphically depicted on the computer (Fig. 1.5). It has a literal translation, too. As Gleick described it in his seminal work *Chaos*, "... a butterfly stirring the air today in Peking can transform storm systems next month in New York."[33] It is also known by the now-familiar chaos term of *sensitive dependence on initial conditions.*

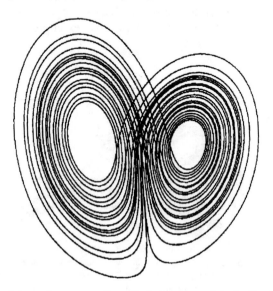

FIG. 1.5. A butterfly attractor. From R. Robertson and A. Combs (1995). Copyright © 1995 by Lawrence Erlbaum Associates. Reprinted by permission.

It means that small perturbations at the outset of a system, or in a chain of events, can have enormous effects on the outcome. In other words, small, almost imperceptible disruptions in one part of a system can have dramatic effects on the system as a whole. Therefore, any small change in the starting point of a system can significantly alter the end point. Hence, any nonperiodic or nonlinear system would be unpredictable because it would be impossible to know all the starting parameters.

Two other chaos terms, *fractals* (or self-similarity) and *bifurcation*, have an integral relationship. *Fractals*, named by the mathematician Rene Mandelbrot, combines the Latin word *frangere*, meaning "to break", with *fraction*, meaning "part of a whole." The whole can be broken down into parts, each of which still resembles the whole. The Koch snowflake (Fig. 1.6) is an example of a simple fractal created by an algorithm that in each successive stage subdivides a triangle.

Fractals can be best illustrated by imagining that you are standing on a cliff looking down at the ocean. Below you, the distant shoreline appears as a jagged pattern. As you climb down the rocks toward the water, there is growing complexity. The rocks along the shoreline appear larger and can more easily be distinguished from one another. Yet, on close examination, the new pattern you observe bears striking similarity to the original one seen from the cliff. Fractals exhibit the characteristic of self-similarity at all levels, across scale size. In other words, the degree of irregularity remains constant across different scale sizes, a kind of regularity in irregularity. However, although there is self-similarity, there is not sameness.

Even our personalities evidence a fractal quality: our handwriting, the clothes we wear, the way we keep our houses or cars, even the way we

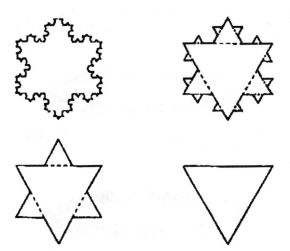

FIG. 1.6. A Koch Snowflake. From M. Butz, L. Chamberlain, and W. McCown (1997). Copyright © 1997 by Wiley. Reprinted by permission.

brush our teeth. A year of life is mirrored in the activity of each day.[34] Fractal images can be seen in trees, their branches, their leaves, and so on. Like nesting Chinese boxes, contained within each box is a small replica of the original. Blood vessels in the body branching out into smaller and smaller capillaries are mirror images of one another, albeit each level on a different scale.

The behavior of individuals, or groups, or other social systems may be similar from "day to day and year to year, or even generation to generation, but no one embodiment in any given cycle or iteration of any given system is precisely like a previous embodiment."[35]

The geometry of the torus represents this self-similar cycle and is akin to first order change, discussed in chapter 4. The figures depicted in Figs. 1.7, 1.8, and 1.9 show how the torus is created by succeeding cycles of similar, but not the same, behavior. I teach a large class each fall entitled Psychological Change and the Spiritual Journey. The format and basic structure of the course (syllabus, requirements, classroom, etc.) remain the same from year to year, but there are slight variations in reading assignments, paper topics, small-group assignments, different students; however, overall, the course has a self-similar quality to each year. Even if I were to attempt to replicate the class exactly, using the same material, my presentation of it would vary from year to year. It would be impossible

FIG. 1.7. A torus after one cycle or one year of class. From R. Robertson and A. Combs (1995). Copyright © 1995 by Lawrence Erlbaum Associates. Reprinted by permission.

FIG. 1.8. A torus after five cycles or years of class. From R. Robertson and A. Combs (1995). Copyright © 1995 by Lawrence Erlbaum Associates. Reprinted by permission.

FIG. 1.9. A torus after 20 cycles or years of class. From R. Robertson and A. Combs (1995). Copyright © 1995 by Lawrence Erlbaum Associates. Reprinted by permission.

to recapture the subtleties or nuances from previous lectures. Using the fractal geometry of the torus to map each year's class yields similar cycles that are loose approximations of one another.

You can see that each year's class is similar, but not the same. This subtle variety, from iteration to iteration, is the essence of human and social interaction, not conformity. It is the slight variety that makes living systems adaptable.

The torus figures also depict first-order change. (The butterfly attractor represents second-order change.) The slight variety evident from iteration to iteration contains the seeds of possible second-order change and significant transformation and under certain conditions, can evolve into a more complex system (Fig. 1.10). Unfortunately, contemporary research tends to ignore these subtle variations, from iteration to iteration, because they are unable to effectively measure them.

Groups exhibit these self-similar patterns. They develop along predictable lines, form norms of behavior, develop leaders, complete tasks, attend to members, and so forth. Yet, no one group is identical, nor can any one session be replicated. Most groups are contained by societal norms of behavior; variations outside of acceptable behavior are usually damped by the group. Over time and with examination of many different group iterations, a fairly similar pattern of group development emerges, such that articles and books can be written about group dynamics. The hidden danger in this practice is that over time, we ignore the subtleties in each group iteration (depicted by the torus) and reduce the three-dimensional dynamic process of groups to a flat two-dimensional form that can more easily fit with the statistics of modern research.[36]

In chapter 3, the concepts of self-similarity and fractals are used to explain group development. Furthermore, recognition of stages, substages,

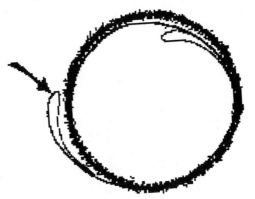

FIG. 1.10. A Poincaré section of a torus after 8,000 iterations. The tongue forming on the outer edge signals the beginning of second-order change. From R. Robertson and A. Combs (1995). Copyright © 1995 by Lawrence Erlbaum Associates. Reprinted by permission.

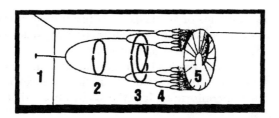

FIG. 1.11. Cascading bifurcations. From R. Robertson and A. Combs (1995). Copyright © 1995 by Lawrence Erlbaum Associates. Reprinted by permission.

and individual group sessions as fractals of the group as a whole provides a foundation for building relationships among these various group levels.

Bifurcation refers to the process of splitting a system into matching parts as the system moves from order into chaos. A *bifurcation point* is the critical value in the system beyond which the system will evolve toward a new state. As the system becomes less stable, passing through more bifurcation points, it has greater difficulty returning to a stable condition. Bifurcation is also referred to as *period doubling*. At each fork or bifurcation point, the system moves farther from order, doubling the amount of time necessary for it to return to its previous steady state. Furthermore, system complexity increases, as do the attractors. The sequence depicted in Fig. 1.11 tracks the increasing complexity of the attractors, the system's movement from order into chaos.

Figure 1.12 depicts system movement from a steady state through a series of bifurcations into chaos. The horizontal lines represent order, the vertical lines, chaos. Each bifurcation (fork) represents two distinct choices available to the system. Like Frost's poem about two paths diverged in a wood, choosing one leads the system in a new and irreversible direction, and that choice depends on which point gets amplified.

At the heart of chaos theory, and its possible application to social systems, lie the assumptions of self-organization, dissipative self-organization, and

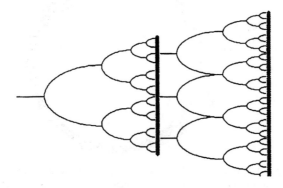

FIG. 1.12. Stability and bifurcation. From M. Butz, L. Chamberlain, and W. McCown (1997). Copyright © 1997 by Wiley. Reprinted by permission.

antichaos.[37] *Self-organization* refers to a process "by which a structure or pattern emerges in an open system without specification from the outside environment."[38] Much of the initial work on self-organizing behavior was done by Jantsch,[39] Prigogine, and Stengers.[40]

Prigogine and Stengers' seminal work sees the universe in terms of a dissipative self-organizing principle. Incompatible with the second law of thermodynamics, or entropy, which augurs a world winding down, the self-organizing principle views the world as progressing from disorder to order. Two oft-cited examples of self-organization, the slime mold and the Beluzov–Zhabotinsky (BZ) reaction, are described to illustrate their theory.

Slime Mold

There are two phases to the slime mold's life (Fig. 1.13). It exists, first, as a single-celled amoeba that leads its own individual life scavenging for food. Yet, when it is deprived of the bacteria it consumes, it radically changes and undergoes a phase transition. It emits (pulses) a chemical signal (cyclic AMP) that attracts other amoebas. In wavelike patterns, thousands of these cells begin to cluster. Then, an internal transformation (Phase 2 transition) takes place, and these individual cells aggregate and become one differentiated animal called a *pseudoplasmodia*. Now, with a head and a body, the pseudoplasmodia migrates along the forest floor to another location where food is present. Then, the spores on the head (fruiting bodies) of the pseudoplasmodia break open, and out come individual amoebas, completing the life cycle.[41]

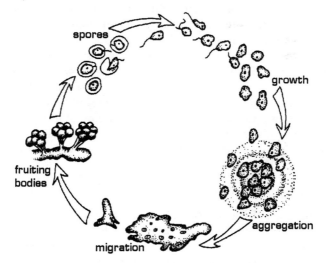

FIG. 1.13. Life cycle of the slime mold. From J. Briggs and F. D. Peat (1989). *Turbulent Mirror: An Illustrated Guide to Chaos Theory and the Science of Wholeness.* Copyright © 1989 by J. Briggs and F. D. Peat. Reprinted by permission of HarperCollins.

Phase-Locking or Entrainment. The rhythmic pulsing of the individual amoebas that facilitates the collective reorganization and the spatial reorganization in the BZ reaction is known as *phase-locking,* or *phase entrainment.* The classic example of entrainment is cuckoo clocks. Hang a number of cuckoo clocks on the wall with their pendulums out of synchronization and after a while, almost magically, the pendulums are synchronized.

> As each pendulum swings, it sends small perturbations through the wall that affect its fellows; it is, in turn affected by similar perturbations from the other clocks. The many independent cuckoo-clock systems "couple" into a larger coordinated clock system; they act as one. Entrainment is the technical term for this tendency to couple into larger wholes.[42]

The phenomenon of entrainment or phase-locking also appears throughout nature: Fireflies flash in harmony; crickets chirp in concert; women living together tend to synchronize menstrual cycles; and, even lovers sleeping together breathe in unison.[43] As Hooper and Teresi noted, "The quintessence of nature's self-organizing principle is consciousness."[44] They cited several examples of how consciousness routinely arranges randomness into patterns. Human memory is replayed as a story, not as an accurate script. The mind arranges stars in the sky into constellations; biological forms are sorted into phyla, genera, and species; chemicals are chronicled in a periodic table; and letters of the alphabet are turned into the great classics.

At all levels of relationship, there is a tendency toward marching to the beat and rhythm of the same drummer. However, nowhere in the schemata of individual actors can emergent collective behavior be explained. The whole is not merely additive, but clearly greater than the sum of its parts. Apparently, when large numbers of cells come together, a critical mass is reached and cooperative collective behavior results. In social systems, smaller numbers of individual actors have been shown to be more amenable to cooperative collective action than larger groups, particularly when group members expect to be together for long periods of time.[45] Biological evolution is thought to occur through the "coupling of independent forms into more efficient and more creatively adapted cooperatives."[46] The Gaia hypothesis[47] proposes that even the biosphere interacts in a coordinated and self-maintaining manner.

Beluzov–Zhabotinski Reaction

Understanding the characteristics of the Beluzov–Zhabotinski reaction also has relevance for social systems, again by illustrating the concept of self-organization.[48] Prior to this discovery, chemical reactions were thought to return to a state of equilibrium. In this experiment, inorganic chemicals are placed in a beaker, dyes are added, and the mixture is continually stirred. A pattern emerges as the chemicals oscillate (periodic attractor) in sudden and discontinuous jumps, between red and blue, forming a kind of chemical clock. These oscillations are temporary. When the solution is poured into a petri dish and left undisturbed, an entirely new organization emerges, one with spatial oscillations. Perturbation in any region of the solution causes instability, which results in the formation of spiral or circular waves. These waves slowly propagate throughout the system with a diverse set of frequencies. *Entrainment* occurs when a faster wave overtakes a slower wave, making it disappear (constructive interference). The equations describing both the BZ reaction and the self-organizing behavior of the slime mold can be applied to other natural phenomena as well.[49]

Where else do we find systems possessing a high degree of naturally generated organization that are highly sensitive to perturbations, switch rapidly from one state to another, and operate spontaneously? Schore responds, "In living systems."[50] The slime mold and BZ reaction serve as possible explanations for the emergence of human social order that arises out of the random, unpredictable behavior of individuals.

This overview of chaos introduces you to this emerging paradigm. Whether, as Loye and Eisler asserted, it is the third wave in science that remains to be determined.[51] However, the commotion that chaos theory created in the natural sciences spilled over into the social sciences. An important bridge between ideas, developed in the natural sciences and those of social scientists, is reflected in Jantsch's creative work *The Self-Organizing Universe*.[52] Other bridge building was done in the areas of synergism,[53] general systems research,[54] urban and transportation systems growth patterns,[55] and social, economic, and political change.[56] However, a direct translation of chaos theory from the natural sciences to the social sciences, without first developing an action-oriented social theory, may be difficult.[57] Furthermore, the qualitative difference between chaos study in the natural sciences and in the social sciences is a normative one. Studying the patterns of rapidly shifting chemical reactions or wildly fluctuating numbers can be done with no consideration of whether the patterns are good or bad or whether newly organized stable states are more healthy than the previous ones. However, when the boundary into the study of living systems is crossed, answers to these ethical questions predominate.[58]

In order to effectively adapt chaos theory into the social sciences, a name change is proposed. It is argued that the word chaos is actually a misnomer when applied to social systems.[59] In actuality, most of us do not live in, nor is our social world dominated by, chaos or constant upheaval. What is present is a combination of order and disorder, linear and nonlinear dynamics. Therefore, it is proposed that transformation theory be adopted as the social science equivalent for chaos theory because it not only subsumes both chaos and order under its broad title, but draws attention to the primary theoretical value that chaos theory holds for social systems. In others words, "the idea of transformation as a process out of or through which order gives way to chaos, and chaos again leads to order."[60]

Chapter 2

Chaos in the Social Sciences

This notion of transformation or chaos is not a foreign idea in the social sciences. Its roots can be traced back to early formulations of dialectic theory found in the ancient Chinese Book of Changes, the early Greek philosophers, and through the works of Hegel, Marx, and Engels. Even Prigogine[1] noted the effect that Comte, Durkheim, and Spenser had on his formulation of dissipative structures.[2]

Pioneering social scientists wrestled with the same questions of change as their counterparts in the natural sciences who were examining the underpinnings of chaos theory.[3] Aspects of group formation or nucleation, solidarity, conformity, and norm formation, articulated by Durkheim[4] and others, find current reflection in Prigogine's notion of cross-catalysis. Durkheim's notion of *anomie* attempts "to describe the psychological effects of the breakdown of norms and social expectations that characterize social chaos states."[5] Even the concept of alienation could be interpreted as a reaction to the constraint of too much order.[6] Other theorists in sociology,[7] psychology,[8] and history[9] grappled with the dynamics of change and non-equilibrium states.

A growing body of psychological literature reveals that chaos theory and its applicability to the social sciences are being explored. Areas in which chaos and nonlinear dynamics have been applied include psychology,[10] psychotherapy,[11] family therapy,[12] memory and cognition,[13] psychoanalysis,[14] multiple personality disorder,[15] schizophrenia,[16] psychiatric disorder,[17] Jungian therapy,[18] posttraumatic stress disorder,[19] education,[20] counseling,[21] society,[22] and social science.[23]

General systems theory built on the early work of Bertalanffy,[24] who drew ideas from cybernetics, is complementary with chaos theory. Although general systems theory addresses both stability and change, it focuses on the former. Conversely, chaos theory explores how systems change.

Lewin's ideas in social psychology have direct correlation to chaos theory. Drawing from his field theory, he expressed the basic transformative notion of social change as a three-step process. First, the current stage is disrupted by some action that perturbs and "unfreezes" it. The second stage is characterized by movement to a new level. Then, finally, a refreezing occurs during the third stage to prevent a return to the previous level.[25] Lewin had notions about nonequilibrium states, even mentioning them, but he never fully developed his ideas.[26]

Although the heuristic value of applying nonlinear dynamics to human systems is debated,[27] recent brain research[28] seems to offer compelling evidence that the processes by which the brain creates memories is very similar to the processes that drive an oscillating chemical reaction (e.g., BZ reaction described in chap. 1).[29] Although "the complexity and irreversible process of growth and pattern development in psychological systems" makes use of nonlinear mathematics impossible, the utility of applying the characteristics and dynamics of self-organizing to psychological systems is beneficial.[30] It is important to remember that "terms that refer to specific and limited ideas in mathematics and physics should not be confused with the broader characteristics of self-organizing psychological systems."[31]

THE SELF-ORGANIZING BRAIN

One study, whose findings went well beyond the initial hypotheses posed when the study began, investigated the manner in which odors are remembered in the brain. Using rats, Freeman[32] not only uncovered how odors are represented in the brain, but he also gained valuable knowledge about the brain connection between nonlinear systems and memory.[33] Initial attempts to consistently measure the carrier wave's frequency and magnitude, both between and within odor exposures, was unsuccessful. When measuring only physical patterns, the EEG was unable to distinguish between odors. In a flash of insight, Freedman placed an array of electrodes in the rat's brain, rather than a single one in the olfactory lobe. He then discovered that each odor could be delineated by the spatial pattern of the amplitude of the wave.[34] The elicitation of the carrier wave, which reflects the collective actions of pools of neurons after exposure to an odor, could be reliably verified.[35] The message that this is peppermint "was not in waveform at all; it was in the spatial pattern of the amplitudes of the waveform."[36]

Freedman further discovered an example of the brain's self-organizing capacity. He found that in learning a new aroma, the brain reorganized previous odors into new spatial structures. Not only was this additional evidence of the brain's self-organizing capacity, it also confirmed another important tenet of self-organization in chemical and biological systems.[37] Freedman found that an odor could be detected in any part of the bulb and rapidly spread throughout the entire bulbar structure. In other words:

> Information necessary for creating a spatiotemporal structure is stored in nascent form throughout the system and in which a perturbation can generate this structure through a wavelike response is a hallmark of self-organization in chemical and biological systems.[38]

This is much like one aspect of the fractal concept discussed earlier, where the whole is contained in each part comprising it.

Freedman's research underscores the dilemma in accurately modeling complex behavior in the neurosciences. The difficulty lies in being able to qualify completely all of the variables or to isolate them from their complex interrelationships. Learning fosters a process of pattern development and growth, making mathematical modeling infeasible. Therefore, in the study of psychological and social systems, the characteristics and dynamics of self-organizing systems are important, not the math.[39]

SYSTEMS AT, NEAR, OR FAR FROM EQUILIBRIUM

Now, let us examine systemic change from the perspective of systems at or near stability and those far from it. This discussion is based on the work of Elkaim, who summarized both the work of von Bertalanffy and Prigogine.[40]

General systems theory was the cornerstone of the family therapy field. Yet, as Elkaim correctly asserted, it mainly applies to systems that are at or near equilibrium. The theory focuses on steady-state systems wherein perturbations are damped; under certain conditions, the system remains within prescribed boundaries. This differs considerably from a system that is far from equilibrium wherein perturbation, under specific conditions, can amplify that system to such a degree that it evolves to a new state. Prigogine's work on dissipative structures focused on these latter systems. In systems that are far from equilibrium once a critical threshold is reached, the fluctuations become amplified such that the system cannot return to its previous or steady position. Instead, it moves toward a new state. Elkaim utilized the life cycle of the slime mold, discussed earlier, to illustrate how a system far from equilibrium changes.

Slime molds are composed of amoebas that live and multiply as unicellular organisms. They feed on bacteria on forest floors. When their food supply disappears, their steady state becomes unstable. At a critical point, the individual amoebas begin to aggregate to form a multicellular structure. The amplifying factor is a chemical signal, sent out by the cells, that facilitates the aggregation. Individual amoebas are attracted (phase-locked) and incorporated into the forming aggregate. The aggregate evolves into a multicellular structure, forms a head and stalk, and moves to a new food supply, whereon the head breaks open, releasing spores that produce new amoebas.

The conclusion Elkaim reached from this example is that once a system passes a critical threshold, in this case, the beginning of aggregation, further fluctuations tend to propel the system toward a new state rather than return it to the previous one.

As systems move farther from equilibrium, becoming less stable, fluctuations begin. Each fluctuation is a potential path or bifurcation point leading toward a new state or organization. Here, chance enters the equation because it is not possible to know, with any certainty, which fluctuation or path will be adequately amplified to push the system in one particular direction or another. Elkaim used the example of termites cited in Prigogine to underscore the role of chance in this process.[41]

African termites build their nests in three steps. They construct pillars, connect the pillars to form arches, and then fill in the spaces between the arches. The process begins as termites scatter small piles of various materials at random. Smells from these materials attract other termites. The stronger the pile's smell, the more termites are attracted to it. Once a pile reaches a critical threshold (stronger smell), it is amplified by other termites depositing more materials, and the pillar rises. Until the critical threshold or critical mass—stronger smell—is reached, it is impossible to tell which of the piles forms the pillars.

The gasoline engine edged out the steam engine because of something that had nothing to do with technological superiority. The steam engine was irreversibly hindered by an outbreak of foot-and-mouth disease among cattle. The outbreak resulted in a law requiring the removal of water troughs that were the primary source of resupply for the steam engine.[42]

In the world of merchandizing and marketing, similar phenomena occur daily. Two recent examples demonstrate how even superior products can lose market share; unforeseen quirks and anomalies often contribute more than marketing does to the success or failure of a particular product.

Competition between the VHS and Beta-Max video recording systems is a good illustration. Both systems were good, yet some suggested the Beta system was superior. However, the VHS system had a ready and available supply of recorders, which gave the machine a slight edge over the Beta-

Max. As people purchased the VHS system, demand for tapes and movies produced on that system increased, setting in motion a process whereby the VHS system reached critical mass and eventually eliminated the Beta-Max system.

Apple Computer, formerly a cutting-edge operation, recently reported a $700 million loss for its current fiscal year. Apple continues to lose market share to the IBM-compatible computers that use a different machine language. In fact, the popular, IBM-compatible Windows format was originally developed by Apple. A decision to not permit cloning of Apple's basic language, combined with the explosion of MS-DOS compatible machines, has led Apple to a point of critical instability, and the company's future existence is now very uncertain.

THE EFFECT OF HISTORY ON SYSTEMS

The critical point that Elkaim[43] drew from his interpretation of Prigogine's work is the influence of history on systems at or near equilibrium and those far from it. As Elkaim asserted, in systems at or near equilibrium, stability is the rule. System behavior follows general laws that are fairly constant and predictable. In addition, stable systems dampen perturbations to maintain their original state. As Elkaim concluded, "The history of the system's fluctuations takes place within the norms of the system."[44]

In comparison, systems far from equilibrium are governed not by general laws, but by the kinds of interactions between elements or by properties of the systems themselves. Such interactions create instabilities that perturb systems toward different and new modes of functioning. Elkaim emphasized that the mode chosen depends on the system's history. However, given the effect of chance, history can influence the future course of a system's evolution, but not determine it. In other words, history circumscribes the choices or fluctuations that systems undergo as they move farther from equilibrium. However, selection of the fluctuation that gets amplified to a bifurcation point by the system is a random process.

Elkaim incorporated these ideas into his work with families; "It is the family's own unique properties and the random amplification of certain of its singularities that will bring it to a new stage in its development."[45] We return to this discussion when we examine group development and how leader interventions might serve to amplify certain group fluctuations over others. Now, let us examine how these principles of self-organization and change are applied to individuals, families, and society. Butz's rendering of Jungian Psychology, Gottman's research on family structures, and Laszlo's theory of societal evolution offer innovative examples.

JUNGIAN PSYCHOLOGY

Several articles appeared identifying parallels between aspects of chaos theory and Jungian Psychology,[46] none better than Butz,[47] who integrated ideas from chaos theory with the psychodynamic developmental theories of Freud,[48] Erickson,[49] and Jung[50] to clarify and extend Jung's concept of the self. Butz defined *chaos*, in human experience, as overwhelming anxiety. This anxiety, which acts as a preconscious gestation period,[51] foreshadows potential psychic growth. Like the periods artists bear before the leap of creative insight occurs, these cycles of intense distress are necessary for psychological growth.

Butz utilized Erickson's[52] developmental model to make his point that turbulence and anxiety are necessary conditions of, in this case, psychosocial evolvement. In Erickson's model,[53] each of the seven stages of development is seen as a crisis point: basic trust versus mistrust, autonomy versus shame and doubt, initiative versus guilt, industry versus inferiority, identity versus role confusion, intimacy versus isolation, and generativity versus stagnation. At each stage or crisis point, the individual seeks resolution of the basic dialectic between individual and cultural needs. The conflict, turbulence, and resolution are the reorganizing ingredients that propel the individual toward ego integrity. As Butz pointed out, although Erickson's model is linear, through the lens of chaos theory, its power increases.

Jung's[54] theory of development centered on the middle and later stages of life. He ceded much of the early stage theory to Freud.[55] For Jung, the ultimate developmental achievement was the attainment of self. Self, he felt, could be realized by moving beyond the false egoic center that is created by the mind. In Jung's theory, the self stands as the true center point between the unconscious and conscious mind (Fig. 2.1). As the

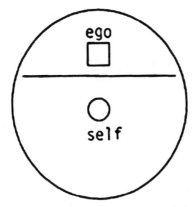

FIG. 2.1. Jung's relationship between self and ego.

center of the personality, the self is firmly grounded and is a point around which all character constellations revolve. Somewhat analogous to Erickson's psychosocial model, Jung characterized the quest for self as a process of separation from collective or societal norms.

However, as Butz pointed out, this midpoint state implies a kind of equilibrium that suggests a closed system and, according to the second law of thermodynamics, eventual entropy. However, Butz argued that Jung offered conflicting evidence on this point. He introduced Jung's notion of *enantiodromia*[56] as a countering force that develops in individuals to compensate for one-sided tendencies. To Butz, this suggested that the psychic is never really able to isolate itself and thus, is subject to the same laws of open systems that apply in chaos theory. So, utilizing this evidence, Butz redefined the self as transitory, rather than static, thus, enabling it to be modeled by chaos theory.

Butz explained that individuals are able, during stable periods in their lives, to achieve a fixed, yet transitory, sense of self. However, these periods remain stable only until the psyche encounters novel material that it is unable to integrate within its current mental configuration. When the mental apparatus is disrupted, chaos ensues, followed by a period where the organism reorganizes at a higher level of complexity. This process seems compatible with that inferred in Freeman's[57] brain research described earlier. As the organism develops higher and higher levels of complexity and adaptation, it alternates between periods of stability and chaos. However, as Butz noted, the chaotic periods are far less frequent than are the stable ones.

Butz, like Freedman, realized one of the fundamental tenets of chaos theory—the self-organizing capacity of living systems. Their work corroborates the earlier research of Prigogine and Stengers[58] in chemistry, Elderidge and Gould[59] in evolutionary theory, and Maturana and Varela[60] in biology, who have confirmed that living systems appear to be able to generate their own new forms "from inner guidelines rather than the imposition of form from outside."[61]

According to Butz, psychic chaos and subsequent self-organization signal a creative gestation period wherein the psyche reorganizes itself to accommodate or integrate novel material. Both Butz and Jung[62] discussed the link between chaos and creativity, recognizing what so many others have, that psychic turbulence is a necessary condition prior to new insight or creation of a new psychic structure. As an artist might struggle with containing chaos in order to create, so, too, must an individual in the throes of psychic upheaval manage chaos while undergoing a transformation. During chaotic periods, the unconscious issues form symbolic images or mandalas. These mandalas, containing symbols of the self, are expressed in a mathematical structure. They appear to be compensatory. Mandalas both express and create order in apposition to ongoing psyche chaos. Butz concluded that:

> These symbolic representations of the transitory self may also act as a container to focus chaotic experience toward an organized state. As a consequence, the mandala or the symbol seem to function as an attractor that brings about order.[63]

What is fascinating about these mandalas are the incredible similarities they have to the fractal images so prevalent in the geometry of chaos.

In counseling, the therapist's task is to create order, or, as Winnicott[64] so aptly entitled, a holding environment that contains the client's overwhelming anxiety. The container soothes the client's anxiety, but does not interrupt or dissipate it, or interfere with the natural psychic reorganization. Furthermore, the therapist does not attempt to order the disorder, but instead validates it as a necessary precondition for change. Therefore, containment, as used here, legitimates chaos and does not control or restrain it. In order to create, artists learn to appreciate the necessity of chaos as a prelude to new insight. As a result, "Creative people tend to be more tolerant of ambiguity in perception than less creative people . . . and prefer chaotic and irregular shapes" to more symmetrical ones.[65]

FAMILY STRUCTURE AND CHANGE

Gottman's research applied chaos theory to the study of family systems and how those systems change.[66] He identified two types of change, regulated and chaotic, that bear resemblance to ideas of first- and second-order change[67] and to the more recent notions of linear and nonlinear transformation. Gottman was less optimistic about the reorganizing potential of chaotic change than were others.[68] However, his ideas and mathematical model for measuring chaotic change are quite notable and worth examining closer.

Gottman compared regulated change to the concept of homeostasis, implying, like the proverbial thermostat, that in family systems, a mechanism exists that regulates and limits behavior. So although the family is not static, it remains relatively stable over time because of this control mechanism that limits a family's ability to deviate greatly from the status quo. A similar mechanism exists in the body, called the *baroreceptor reflex*, which operates the cardiovascular system and, among other things, keeps blood pressure fairly constant. These regulated systems are quasiperiodic, meaning they go up and down over time. One other feature of regulated systems is the relation between input and output. Small changes in input produce small changes in output. This differs sharply from the deviation amplifying systems characteristic of chaotic systems, wherein small deviations in input can result in immense changes in output.

The often used predator-and-prey model provides an excellent example of regulated change. An imaginary ecosystem is depicted on a graph (Fig. 2.2), referred to as state or phase space. In this imaginary world, the population of big fish (predator) at any time, t, runs along the y-axis. The population of little fish (prey) at time t hops along the x-axis. Charting the relation between big and little fish produces four areas on the graph.

The first area (A) depicts the ecosystems relation between small populations of big fish and little fish. Big fish decrease due to lack of food, whereas little fish increase because of the lack of predators. The second area (B) depicts a relation between few big fish and many little fish. Predictably, both populations increase. The third area (C) contains large populations of big fish and little fish. Under these circumstances, the big fish population increases, whereas the little fish decrease. In the final area (D), there are many big fish and few little fish. Thus, both populations decline.

Plotting more data points the vectors round into a circle (Fig. 2.3), creating a periodic system with variability in both big and little fish populations. As Gottman pointed out, a set of concentric circles (Fig. 2.4) represents all possible predator–prey relationships. Thus, as Gottman asserted, every trajectory of the "system is periodic, with periodic variability in the population" of big and little fish.[69]

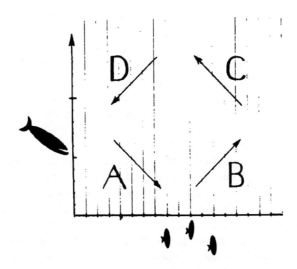

FIG. 2.2.   Phase of a stable ecosystem. From J. M. Gottman (1991). Copyright © 1991 by Lawrence Erlbaum Associates. Reprinted by permission.

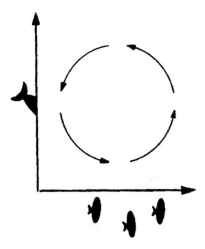

FIG. 2.3. A stable ecosystem represented by periodic motion in a circle. From J. M. Gottman (1991). Copyright © 1991 by Lawrence Erlbaum Associates. Reprinted by permission.

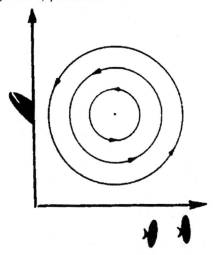

FIG. 2.4. Concentric circles representing all possible states of the stable ecosystem. From J. M. Gottman (1991). Copyright © 1991 by Lawrence Erlbaum Associates. Reprinted by permission.

Gottman refined his notions of regulated change when applying them to families. He extracted two terms from physiology that metaphorically represented, for him, the homeostatic dialectic inherent in family systems. He proposed that the terms *catabolic* (energy conserving) and *anabolic* (energy consuming) be used to describe the energetic social interactions among family members and between families and their environment. He concluded that family processes are in balance if energy conservation and expense are roughly equivalent.

Unlike regulated change, which is expected and even predictable, chaotic change is unpredictable. There is no set point at which energy is considered to be balanced. Therefore, homeostasis is violated. Of course, what makes the behavior of chaotic systems so unpredictable is that small changes in initial conditions can result in gigantic variations in the system's trajectory. The net result is that we cannot say with any certainty where the system will end up. In regulated systems (e.g., big and little fish), once the initial conditions are determined, the end result is known.

Gottman defined an attractor in family systems as a fixed point, that is, a balance between catabolic and anabolic processes or positive and negative affect. In other words, *fixed points* for Gottman were defined "as all those things that can be referred to by family members to resolve conflict in such a way that family cohesion increases after conflict."[70] Fixed points connote a sense of "we-ness." Examples of such notions of we-ness include:

> the implicit marital contract, a shared religious or cultural viewpoint, and agreed-upon dominance hierarchy, the family's belief system, shared values, shared goals, shared memories, a shared viewpoint of reality . . . the family's stories, heroes, demons, and myths, and expectations the parents have of their marriage, the family's rules, rituals, beliefs about people, good and evil, etc.[71]

These fixed points provide the foundation for ordered change. Without them, the family faces a break-up every time there is conflict because fixed points are the glue (cohesion), or sense of "we-ness," that holds the family together during stressful times. However, should the family's fixed points become endangered (e.g., extramarital affair, terminal illness), chaotic change may result. Without re-establishing the fixed points, the family may remain in a state of disorder.[72]

As mentioned, Gottman translated catabolic and anabolic functions in the family into positive and negative affect. Catabolic or energy-expanding affects are anger, fear, sadness, or any combination of the three. Conversely, humor, amusement, interest, and affection represent anabolic or energy-conserving emotions. Therefore, a regulated system maintains a balance (4:1 ratio) between positive and negative emotions. This means that positive affect should exceed negative affect by at least a ratio of 4 to 1. The 4-to-1 ratio was established in an earlier research study.[73]

In another part of his work, Gottman described the mechanisms by which families deregulate the balance between catabolic and anabolic processes, resulting in chaotic behavior. Using a case study, he plotted the negative affect, over time, of a marital couple in conflict. His interest in this data was not whether negative affect was high in dissatisfied marriages, but whether the system was regulated and stable or chaotic. He computed *interevent intervals*, or the time between display of negative affect, during a conflict discussion of this unhappy couple. In the scatter diagram, a state space is created wherein $x(sub\ i) + (sub\ 1)$ versus $x(sub\ j)$. Where this first-order *autocorrelation coefficient* (a correlation of a series of data points—as in time series data—with itself) differed from traditional research methods and took into account chaos theory is when Gottman connected the dots (Fig. 2.5). Thus, Point 1 or $(x\ sub\ 1,\ x\ sub\ 2)$ was connected to the next point of $(x\ sub\ 2,\ x\ sub\ 3)$ and so on. The scatterplot revealed a tendency for this system to wander toward the tip of the created triangle.

In Fig. 2.6, Gottman extended Fig. 2.5 into a three-dimensional trajectory. The results depict an unstable system in decline (but it has shape). It is dissipative, winding down like the pendulum. Over time, the intervals

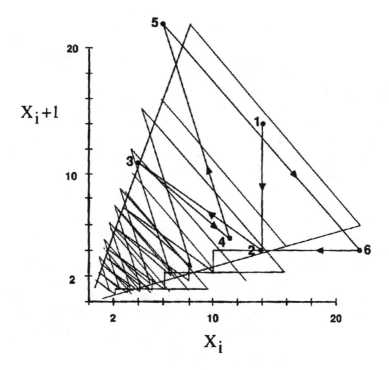

FIG. 2.5. A scatterplot of interevent times with consecutive points connected. From J. M. Gottman (1991). Copyright © 1991 by Lawrence Erlbaum Associates. Reprinted by permission.

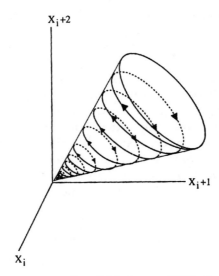

FIG. 2.6. Runaway system. From J. M. Gottman (1991). Copyright © 1991
by Lawrence Erlbaum Associates. Reprinted by permission.

between negative responses are shorter and indicative of a runaway system.
He concluded that such turbulence leads to disastrous results.

Gottman observed that the system he studied—or any, for that matter—
may be part of some larger framework or subject to some macrolevel
regulation: a basic tenet of General Systems theory. However, as stated
earlier, Gottman is less optimistic about the system's adaptive ability than
are others.[74] Gottman concluded that, during regulated or orderly change,
a system is capable of subsequent adjustment and assimilation, unlike the
unpredictability inherent in unregulated change that results from chaotic
systems.

Others view chaos as a means of generating diverse and novel solutions
to complex situations.[75] The key in each situation is the threshold behavior
that occurs as a system moves from order into chaos. This critical, or
in-between, period presents an opportunity to perhaps influence the system's trajectory.[76] Two other useful ideas that can be gleaned from
Gottman's creative work are his notions of fixed and set points. A *set point*
refers to the energy balance between catabolic and anabolic processes.
Fixed points promote a sense of "we-ness" that stabilizes a system so that
conflict can be resolved and cohesion increased. Chaotic systems have no
set points because they are in transition.

Before we apply the principles of self-organizing systems to small group
behavior, let us look at Laszlo's innovative theory, which incorporated
many of the tenets from chaos theory into formulations about evolutionary
change. His ideas provide yet another view of systemic change.

LASZLO'S SOCIETAL EVOLUTION

Laszlo's ideas about societal evolution were framed by concepts from the natural sciences. Drawing on his understanding of chaos theory, Laszlo proposed that both natural and social systems are subject to the same general laws of change or evolution. In societal systems, these laws are influenced and shaped by the beliefs, mores, values, and habits of the human beings who comprise them. Therefore, the laws do not prescribe outcome, but simply set the rules and limits of the game. In addition, social systems are not biologically determined. Instead, they evolve and persist in the multilevel milieu of other systems in the biosphere. Social systems persist as human beings cycle through them, being born, growing, and dying. Accordingly, social systems are not the sum of the human beings who comprise them, but have functions and attributes that allow them to maintain their congruity and evolve independently of human life cycles. Society is self-organizing and has the survival capacity to adapt, evolve, and change to alternative structures if necessary.

> Society evolves through convergence to progressively higher organizational levels. As the flow of people, information, energy, and goods intensify, they transcend the formal boundaries in the social systems. The catalytic cycles that maintain the system in its environment encounter similar cycles in the intersocial milieu and interact with them. In time the cycles achieve coordination as intersocietal hypercycles. Thus neighboring tribes and villages converge into ethnic communities or integrated states, these in turn become the colonies, departments, provinces, states, cantons, or regions of larger empires and ultimately of nation-states. When empires are stripped of their far-flung territories and overseas colonies, the capital regions and the liberated states are open to new forms of convergence among themselves.[77]

The convergence Laszlo described is an example of entrainment or phase-locking. What he described is evident in several circumstances today. In response to the emerging global marketplace, economic organizations (like NAFTA and GATT) were created to supersede national boundaries. The former Soviet Union has undergone significant transition and upheaval since its disincorporation. With the new emerging world order, old world organizations, such as NATO and the United Nations, are under pressure to adapt. It can even be argued that the United States has lost its mooring, and recent rapid swings in the social and political structure signal a pending transformation.

Laszlo's chart (Fig. 2.7) depicts the stages of evolution in modern society. The horizontal axis, time, pictures the evolution of society from the hunting–gathering era to the postindustrial period. The rapidity of contemporary change is reflected when one considers that the hunting–gathering

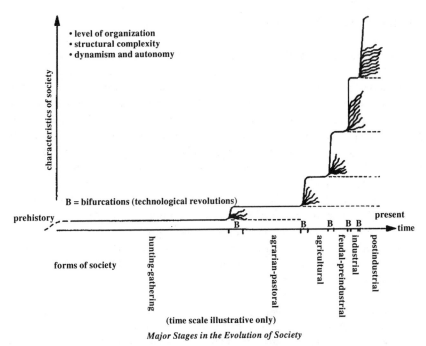

FIG. 2.7. Major stage in evolution of society. From E. Laszlo (1987). Copyright © 1987 by Ervin Laszlo. Reprinted by permission.

period exceeds all contemporary stages from the agrarian revolution through present times. Societal evolution is speeding up.

Characteristics of society are depicted in the vertical axis. In Laszlo's model, evolution leads to greater structural complexity, higher levels of organization, and increases in dynamism and autonomy. The driving force in Laszlo's model is technology. In short:

> Nomadic hunting–gathering tribes domesticate plants and animals and transform into settled agrarian–pastoral societies; agrarian–pastoral societies evolve such technologies as irrigation and crop rotation and transform into agricultural ones; agricultural societies develop handicrafts and simple manufacturing technologies and thus transform into industrial societies; and industrial societies, under the impact of new, mainly information- and communication-oriented technologies, evolve into postindustrial societies.[78]

It is tempting to argue, here, the advantages and disadvantages of technological innovations; however, Laszlo's notions of change are what we are after. Several aspects of his model are applicable for understanding group change. First, change is progressive, not linear. Movement is uneven. There are sudden leaps forward and periods of regression, but overall, there is direction. Second, complexity increases at each stage of evolution. Third, the activity or energy of each stage, as well as its autonomy, is increased.

In Laszlo's model, societal stability is explained by *autopoiesis*: a system's ability to renew itself. Laszlo invoked systems thinking when he noted that societies are dynamic and homeostatic because they oscillate around certain norms. Societies sustain their stability through reproduction and regeneration by insuring adequate availability of natural resources, energy, information, money, raw materials, food, and people. Societal stability is further maintained by the system's ability to absorb and dampen small perturbations (first-order change).

However, as Laszlo claimed, societies not only renew themselves, but are also capable of significant and second-order change. They can be perturbed to instability by major events, among them war, economic upheaval, and technological revolutions. For example, robotics had a major impact on the industrial world, causing displacement of tens of thousands of workers. The computer revolution continues to excite the stability of contemporary society, and ethnic wars in Eastern Europe are redefining national boundaries. Such events illustrate how previously stable systems can be pushed toward disequilibrium. When societies cannot maintain themselves and are perturbed to the point of critical instability, than anarchy or chaos may ensue, leading to second-order change. The resulting bifurcations may lead to the emergence of new social, political, and economic structures, or the society may be absorbed into a more powerful system.

A crucial point Laszlo made about these bifurcations and the resulting new social order is that they are indeterminate in regard to historical direction. They can either be progressive or regressive. However, the sum of these transformations, as he illustrated on the vertical axis of Fig. 2.7, is progressive. Thus, although societal evolution is not predictable or predetermined, it has an overall evolutionary pattern. Here, Laszlo illustrated one of the basic tenets of chaos theory. Although societal evolution is unpredictable (locally), over time, an overall pattern (globally) or direction emerges.

As technological perturbations occur in society, certain innovations expand more rapidly than others. The wavy lines in Fig. 2.7 represent competing technological bifurcations. Eventually, one will be amplified to the extinction of the others, and that technological advance will organize the next societal level. Examples cited earlier of the steam engine and VHS system suggest that factors other than technological improvement may also influence amplification.

In Laszlo's model, technological advances bring about societal change. Dominant societal paradigms are perturbed by new technologies. As certain technologies are amplified (grow), the system is perturbed toward instability. First-order changes can dampen these perturbations by incremental adaptations. However, if these incremental adaptations are insufficient to dissipate the mounting energy, the system reaches critical instability, chaos ensues, and the system leaps (bifurcates) to a new level of organization. Bifurcations result in second-order change and are discontinuous. Change that results from a single bifurcation may not advance evolution (as defined by the characteristics describing the vertical axis in Fig. 2.7), but the sum of bifurcations or changes do.

In the remaining chapters, ideas generated from chaos theory that were introduced and discussed in this chapter are used to describe how groups change and develop. In the next chapter, a model of group development is proposed that provides the framework for this discussion. In chapter 4, group change is fully explored. Discontinuous change, self-organization, bifurcation, and fractals are translated to group language and applied to the proposed model of group development proposed in chapter 3. Chapters 5 and 6 detail the skills necessary to facilitate groups in transition. Choosing and constructively amplifying group bifurcations are explored. Recognizing and reacting to transition opportunities are discussed.

Chapter 3

Group Stage Model

Group work proliferated during a period from 1950 to 1975 when, after World War II, the benefits of group therapy were widely touted. Those small groups were variously referred to as encounter groups, t- groups, human relations groups, laboratory training groups, and their many derivatives. Rogers described the salient characteristics of these groups.[1]

The group is, in almost every case, small (from 8 to 18 members). It is relatively unstructured, choosing its own goals, norms, and activities. Most of the time, if not always, there is some cognitive input. The leader's primary responsibility is to facilitate the communication of both feelings and thoughts on the part of group members. The main focus is on the here-and-now personal interaction of all participants.[2]

During those years, numerous theories emerged proposing models of development that depicted group growth as moving through a series of discrete stages. The number of stages ranged from 3[3] to 13[4] depending upon the specificity with which each stage was articulated. In fact, most theories could be represented in any number of stages by collapsing or expanding the distinctions made among them.

Since the mid-1970s, few innovative stage theories have emerged. Group work has moved into the mainstream of treatment modalities and, thus, earlier assumptions have been reified as fact. As a result, over the years the distinguishing characteristics of each of these earlier theories have eroded into a nondescript model of group growth. Although there is disagreement as to the exact number of stages groups pass through, there is nearly unanimous agreement that groups do develop sequentially. Only one theorist argues against a clear-cut pattern in the process of group maturation, but his is definitely a minority voice.[5] As a result of this unanimity, a generic five-stage model emerges, popularized in contemporary textbooks, that represents the distillation of these theories. Reified as fact by these textbooks, this generic model becomes the lens through which groups are viewed. Contemporary group practitioners adapt their work to fit their understanding of this dominant stage model.

Rennan[6] reviewed 50 articles on group development and from them selected 16 that best exemplified the range of theories in the literature.[7] She discovered a "remarkable similarity" among the theories and concurred that most models of group development could be distilled into a five-stage model. Equally important was her finding that, without exception, each theory suggested or implied that groups advanced linearly, in a unidirectional fashion. "Each group displays a unique collective personality and a normative patterning all its own. Yet diverse groupings also conform to certain universal principles inherent in joining, forming, and maintaining collective enclaves."[8] Rennan summarized the five predominant stages she found in her review.

COMMON STAGES

An orientation or forming stage was common in each theory. This stage was characterized by high anxiety, a search for meaning and structure, and dependence on the leader.

A single conflict stage was the norm,[9] although three theorists did expand it into two stages. Power and control issues characterized this stage. Testing boundaries and challenging the leader were common attributes. Member frustration at unmet expectations was usually followed by anger directed at the leader. In many of the models, a norming stage was next.[10] It was depicted as an oasis following the turmoil of the conflict stage. Harmony and increased cohesion marked this period. The majority of models described a distinct work stage during which the group actualized its goals. This stage was characterized by authenticity and mutuality among the members. Termination was the final stage delineated in most models.[11] This period was characterized by members validating their group experiences while working to emotionally and physically separate from the group.

My recent work on the characteristics of both regressive and generative groups, discussed in chapters 9 and 10, suggests that three facets of this common model have been inadequately explained.[12] First, in the majority of earlier theories, limited attention is given to the conflict period, and even less to the difficulty groups experience when trying to negotiate this stage. Some theorists even neglected to mention it. The difficulty groups experience in addressing and resolving conflict appears to be more commonplace than these theories suggest because many groups failed to advance beyond the conflict stage of development. Second, the implicit notion that group stages evolve in a linear and unidirectional fashion may be misleading. Third, previous theories never adequately accounted for the winnowing process that groups undergo, the shaping and forging of individual identities into a collective entity. By training I was wedded to the traditional model and tried to fit my observations into it, but there were always loose ends. It was not until I discovered the work of Arthur Young that I had a framework for my ideas.

In this chapter, an alternative model of group development is proposed that addresses these three factors. The model integrates old ideas and new concepts from chaos theory with the work of Arthur Young. It emphasizes the importance of conflict in group development and recognizes that group growth, although progressive, is neither linear or unidirectional. Particular attention is paid to how groups change, evolve, and mature.

YOUNG'S "THEORY OF PROCESS"

Young is the inventor of the Bell helicopter. He is an eccentric with varied interests, one of which is his search for a universal theory of evolution. His book, *The Reflexive Universe*, details his evolutionary "Theory of Process."[13] A brief summary is presented here.

Young proposed a seven-stage model of evolution, beginning with light, building to particles, atoms, molecules, plants, animals, and ending with humans (Fig. 3.1). His evolutionary theory formed an arc divided into two phases, a descent and an ascent. This image quickly conjures up mythical, philosophical, and religious associations, yet Young used contemporary physics to support his proposition. The arc represents the process that evolution undergoes as it moves progressively from complete freedom of movement through a series of stages that constrain it into permanence, at which point it is propelled upward again to complete freedom.

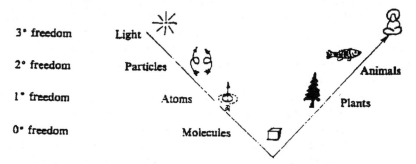

FIG. 3.1. Young's seven-stage model of evolution. From A. Young (1976). Copyright © 1976 by Robert Briggs. Reprinted by permission.

In his model, the descent characterizes a fall into permanence. From the initial freedom of movement of light, evolution moves through a series of three stages of increasing constraint, culminating in the inertness that characterizes molecules found in minerals. At the bottom of the arc is the stillpoint, the liminal area at which the descent or fall into determinism can be reversed by collecting energy and creating organization necessary for the next stage. Although all molecules can occur as crystals, Young acknowledged that not all are constrained into inert objects. Some molecules, during the ascent phase, become organized and evolve into higher forms: plants, animals, and human beings. These higher forms represent a return to freedom of movement. If the timing is correct, the ascent back up toward freedom can occur. Young's discussion of how molecules build energy and move against entropy hinted at Prigogine's notion of dissipative structures.

Timing, for Young, means the correctly timed control of force. The choice of timing is the only hidden freedom left to the molecule in stage four. It enables energy to be collected and used to create the organization in stage five. This correct use of timing is learned during the descent and is a necessary condition prior to advanced stages of evolution. It explains why evolution in Young's model is indirect, a descent, which provides time for learning, followed by an ascent. A similar notion of the proper use of timing by the group leader is discussed in chapter 5.

According to Young, what propels matter through this arc is the action of light. Young made a long and interesting argument for light as first cause. He argued that light is purposive, in part, based on the principle of least action, which asserts that light always follows the path that takes the shortest time. The second argument he proposed is that action occurs in wholes. The whole cannot function when divided into parts. The argument is that parts are derived from the whole, not vice versa. Hence, Young emphasized that the whole exists before the parts. Therefore, Young formulizes, light = quanta of action = whole = first cause. His argument for this assertion is compelling. In viewing light as purposive, he also raised the notion of teleology, an idea I weave into the spiritual discussion of groups.

In sum, there are four ideas in Young's theory, depicted in the arc, that are utilized in constructing a model of group development. First, the arc is symmetrical. Young quantified complementary stages and correlates sides of the arc by counting the axes of symmetry and, conversely, the degrees of freedom at each level.

The first and seventh stages have complete freedom of movement and are asymmetrical. Light has total freedom, as does the seventh stage. In Young's model, it is the highest level of existence and therefore cannot be defined. According to Young, this stage includes, but is not limited to, humans. He pointed out differences in the sides of the human face, right- and left-handedness, and left versus right brain functions as examples of asymmetry.

FIG. 3.2. Level two, stage six, bilateral symmetry. From A. Young (1976). Copyright © 1976 by Robert Briggs. Reprinted by permission.

Stages two and six each have two degrees of freedom and one axis of symmetry. Animals have similar right and left sides, but differ front and rear, and top and bottom (Fig. 3.2). This is known as bilateral symmetry. They have two degrees of freedom and move about two-dimensionally on the Earth's surface. Likewise, according to Heisenberg's uncertainty principle, particles retain two degrees of freedom: position and movement (Fig. 3.3). Bilateral symmetry is also expressed by what is referred to as chirality, or handedness, which characterizes particle reactions.

Stages three and five retain only one degree of freedom and exhibit two axes of symmetry. The tops of plants differ from their roots, but both right and left sides, fronts and backs, are similar (Fig. 3.4). This is known as radical, or cylindrical, symmetry. The plant has one degree of freedom, growing only vertically. Electrons traveling around a central nucleus give atoms their radical symmetry (Fig. 3.5). Atoms maintain their one degree of movement by the unpredictability of their energy state. They can either absorb or release energy.

In stage four, crystals have complete symmetry (Fig. 3.6). Molecules comprising them are arranged tediously in rows, columns, and layers. There is no freedom of movement.

Second, movement on the left side of the arc is random, whereas movement on the right side is voluntary. Life forms on the right have acquired the ability to move intentionally. The movement of light, nuclear particles, and atoms is random. Conversely, plants release and store energy voluntarily, and animals and humans navigate at will.

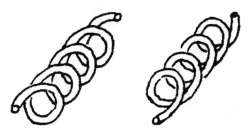

FIG. 3.3. Level two, stage two, bilateral symmetry. From *The Reflexive Universe* by A. Young (1976). Copyright © 1976 by Robert Briggs. Reprinted by permission.

FIG. 3.4.   Level three, stage five, radical or cylindrical symmetry. From *The Reflexive Universe* by A. Young (1976). Copyright © 1976 by Robert Briggs. Reprinted by permission.

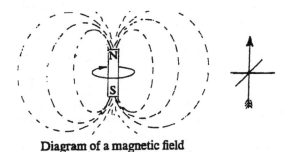

Diagram of a magnetic field

FIG. 3.5.   Level three, stage three, atoms. From *The Reflexive Universe* by A. Young (1976). Copyright © 1976 by Robert Briggs. Reprinted by permission.

Third, at each stage, matter is transformed into a higher level of organization. Each stage is both cumulative and additive. New stages assimilate old ones and add something to them.

Fourth, Young's model is holographic. Each stage is a holon, meaning both part and whole. Particles, atoms, molecules are not only separate entities, but expressions of the whole at successive stages of evolution.

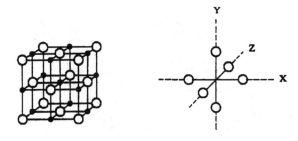

FIG. 3.6.   Level four, stage four, molecules. From *The Reflexive Universe* by A. Young (1976). Copyright © 1976 by Robert Briggs. Reprinted by permission.

Suggested here is the notion expressed in light as first cause, that the whole exists prior to the parts. In the group model, this idea is further developed.

This summary is only a reflection of the breadth and depth of the ideas found in Young's work. He made countless arguments and cited many examples in support of his positions, all of which are beyond the scope of this chapter. Read his thought-provoking book. What is important is that the ideas expressed in his arc (Fig. 3.1) provide an innovative framework on which to construct an alternative model of group development.

In small groups, individuals come together, create a purpose, and forge a collective identity. Initially in that process, individuality is constrained as a group identity forms. The descent represents the collective forging process. The vertex depicts the crucial conflict stage. This is the turning point in groups where responsibility is shifted from the leader to the members. Once a strong bond is established, responsibility assumed, and a group identity emerges, individuality can be reclaimed, asserted, and expressed. The ascent signifies that reclamation process. In a few pages, the model will be fully developed.

OLD IDEAS REEXAMINED

Ideas regarding the winnowing or conforming process in groups can be found in some of the studies reviewed by Rennan. In the beginning stages, some theorists describe a pressure to conform to an emerging, collective identity that results in some members withholding aspects of the self. This period is characterized as "being in or being out."[14] Efforts are made to find commonality with other group members or risk alienation. Members feel pressure to relinquish their individuality and seek commonalties with others as a sign of commitment to the group.[15]

Confinement is another term used to describe this beginning period.[16] Members abdicate responsibility for the group and deskill themselves in deference to the leader.[17] Individuals experience a loss of confidence.[18] Personal boundaries are loosened as members give up preconceived notions of the group and leader; thus they become intellectually and emotionally vulnerable.[19] During this initial phase, it is apparent that members contribute more to the group than they receive from it.

In the middle phase of development, the group feels totally constrained and revolts. Members have made sacrifices for the group and have not had their expectations met. Frustration is high; earlier differences, suppressed for the sake of harmony, reemerge. Unrealistic hopes for the group and the leader give way to reality. Group members feel less pressure to conform, and they begin to participate more authentically. Conflict plays a central role.[20] Resolution of conflict signals the shift in power from the leader to the members. The expression and subsequent resolution of the power issues provide the thrust that propels the group upward.

In later stages, it is suggested, members not only recover what they relinquished earlier, but reap rewards that exceed earlier investments. As the group resolves conflict, members regain a sense of self within the group.[21] Mutuality and increased trust allow the group to achieve new levels of intimacy and understanding.[22] The group gains energy, and freedom is frequently mentioned as a characteristic describing these stages.[23]

These ideas gleaned from the literature reviewed by Rennan hint at a model of development in which individual loss and collective gain play a significant role. Additionally, the conflict stage, which signals the shift in responsibility for the group from the leader to the members, can be conceptualized as the stillpoint that marks this transition. Visualizing a framework to account for these ideas leads us back to Young.

GROUP MODEL

The Arc

The proposed model forms an arc and divides the life span of a group into seven potential stages: (a) Preforming, (b) Unity, (c) Disunity, (d) Conflict/Confrontation, (e) Disharmony, (f) Harmony, and (g) Performing (Fig. 3.7). Seven stages are necessary to adequately and fully portray potential group development. Three stages describe the conflict period

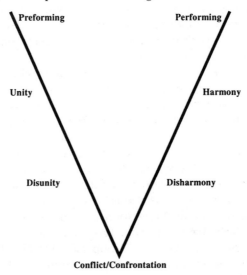

FIG. 3.7.   Group stages arc.

(Disunity, Conflict/Confrontation, Disharmony) to underscore its importance in the overall process. The stages are arrayed on both sides of the arc and joined at the vertex by the Conflict/Confrontation stage. In naming stages, an effort was made to provide accurate descriptors that best characterize the events that comprise them.

Principles of the Arc

Several principles define the arc. The left side of the arc represents the descent in which individuals are conjoined to form a group. The right side, or ascent, depicts group members as an emerging collective force. The arc is symmetrical. Each stage and its counterpart on the opposite side of the arc are intended to reflect one another. What members relinquish in each stage during the descent, they regain in the corresponding stage of the ascent.

Movement through stages in the descent is experienced by members as random. During the ascent, members have voluntary control of their movement. Progress through the stages is uneven as the group advances and retreats, but overall group movement is progressive. Transitions between stages occur by discontinuous leaps, and these nonlinear transformations are disorderly.

Each succeeding stage in the arc represents a higher level of organization. Each stage subsumes the previous one and adds something to it. Successful mastery of each stage is necessary for the group to actualize its full potential.

The arc is a fractal, reflected at every scale of the group, albeit at smaller and smaller levels. Each stage has seven substages, arrayed on both sides of an arc. Every session has seven potential phases, and every interaction could potentially be detailed into seven parts.

Proposed Stages

Preforming. Stage One, the Preforming stage, takes place prior to the first group meeting, as prospective members commit to the group. During this phase, future success is predicated on factors such as group size, member characteristics, and setting.[24] In this period, the group leader organizes the group, determines the structure, and locates an appropriate setting.

Each prospective member has unique expectations for the group and the leader. During the descent, these expectations must be relinquished as members' perceptions give way to reality. In this stage, the group exists as an assemblage of unconscious bonds that form a group mind in which the group exists as potential. All future possibilities are present. Sullivan, the interpersonal theorist, proposed a similar idea regarding individual therapy when he suggested that all the major life themes and issues that comprise the client's life were present in microcosm in the first and every subsequent therapy session.[25] This idea is elaborated at the end of this chapter.

In the most optimum of circumstances, the group's full potential is actualized by the last stage. Very few groups achieve generativity, but it exists to be realized as the group makes its transformative journey through the arc.

During the last stage in the arc, full group actualization occurs as spiritual or transpersonal growth, but is not limited to the collective. Individual members may experience epiphanies within the group. Many meaningful experiences in groups occur through encounters with others. Whether or not those opportunities are actualized does depend, of course, on the group and its members.

Unity. Unity might appear to be a misnomer in characterizing this stage. However, although group members experience considerable anxiety, it is often suppressed and masked. Outwardly, members evidence a conformity, albeit superficial, as the group begins its journey. Members feel compelled to invest in the group before they feel any real connection to it. This stage is in marked contrast to its counterpart, the Harmony stage, wherein members experience a deep inner connection with one another.

When the first meeting begins, members are apprehensive. Issues of safety and trust are the focus as members work to establish norms of behavior that make the group secure. Members experience the group as ambiguous. A predominant feeling in this stage is anxiety. Although it is often masked, the presence of anxiety influences the direction of early group interactions. Members subgroup and project responsibility for action onto others as a way of moderating the anxiety. Beginning group members have not yet established the social conventions by which they will interact with one another; until they do, the atmosphere remains tense.

Members rely on the leader for direction. She represents a powerful figure onto which omnipotent fantasies are projected. In addition to seeking structure from the leader, group members want her approval and acceptance. Dependency issues are present but, as yet, not fully realized.

With apologies to Pirandello, I often label this period "Eight Characters in Search of a Script." Members are unsure of this new environment, so their earliest interactions are focused on defining their group role and those of the other members. This presents itself in many ways, but often groups begin with introductions, where each member gives a brief autobiographical statement. The content of the opening disclosure is dictated, to a large extent, by what the first speaker says. Remember, there are, as yet, no guidelines for group members to follow, so each disclosure and subsequent interaction is carefully monitored for clues to acceptable behavior. In turn, each member usually discloses a similar amount. The content of these initial revelations are usually name, chosen work, and reason for joining the group.

This outwardly simple process of members self-disclosing, in turn, is the basic pattern by which a group develops, and is repeated throughout its lifetime, leading to greater levels of trust and subsequent safety, as members, with each round of disclosure, reveal more and more about themselves. Safety and self-disclosure are inextricably linked: The safer members feel, the more likely they are to open themselves to the group and, in turn, to the feedback of others.

Safety needs occupy the work of the forming group. Members need to feel secure and free from anxiety and disorder. These needs are similar to those of young children, who have strong desires for structure and routine in their lives. New situations are disconcerting, and the child often clings to the parents in such situations. Although adults have learned to inhibit childlike responses to fearful situations, they are still very much in need of a safe, predictable environment.

Without any structure, members experience too much freedom. This leads to ambiguity and increases anxiety. To reduce anxiety, members minimize their differences, temporarily relinquish their individuality, and conform for the sake of unity. Individuals refrain from showing their real selves. Conformity reduces individual expression, freedom, and movement in the group. During this period, group members are concerned with survival, producing strong dependency needs on the leader, who is expected to provide structure, order, and limits.

Members may also respond to the ambiguity and anxiety by leaving the group and not returning. To alleviate anxiety, much of the group focus is spent discussing past issues and avoiding the present. Members also utilize past group experiences as a means of explaining and understanding the present one.

Conflict Stages

Possibly the least understood area of group work is the role that conflict plays in developing groups. To underscore its importance and ensure understanding of its dominant role in determining the success or failure of groups, three of the seven stages address group conflict. Stage Three, Disunity, depicts the increasing frustration and expression of indirect anger among group members. Stage Four, Conflict/Confrontation, delineates the members' direct challenge to the leader for control of the group. Stage Five, Disharmony, represents intermember conflict as the group comes to terms with its differences and diversity.

The Conflict stages are a critical period in a group's development. The expression of conflict and its resulting resolution provides a bridge between the superficial conversations of the first stage, Unity, and the direct expression of feelings in the Harmony and Performing stages. "Without working through this phase and establishing appropriate norms of behavior only a superficial level of cohesiveness can develop."[26]

Too often, conflict has been associated with negative experiences, and individuals have learned to avoid it. Thus, issues of conflict and confrontation are frequently viewed as detrimental to the development of healthy relationships within groups, as well as destructive to the growth of the group itself. This negative view of conflict may have developed because of its long association with aggression.[27] Although anger that exceeds the tolerance of group members may be harmful, its suppression will manifest itself in destructive ways.

In fact, conflict brings "drama, excitement, change and development to human life and societies."[28] It provides the stimulation and potential for group growth.[29] "It is through conflict that existing norms and practices are challenged and changed, and through conflict that we are frequently most creative and innovative."[30]

Conflict is generated primarily from two sources within the group: intermember differences and frustration with the leadership. Conflict is inherent in any relationship. The friction is generated from the differences that naturally exist between people. In groups, working through conflict can help members construct an agreed-on working environment where differences can be understood and accepted.

There is a mistaken belief that conflict and its emotional manifestation of anger are nonexistent in healthy relationships. Quite the opposite is true. In fact, it is the suppression of conflict that deadens relationships. However, before member differences can be addressed, a norm for the expression and resolution of conflict must be established.

Anger is initially directed at the leader for two reasons. First, she is held responsible for the group. Frustration with her failure to meet all expectations makes her a likely target. Second, the group must witness the expression and resolution of anger for norms to be established. With no fixed norms, confrontation with the leader takes place in a vacuum. Unlike members, who appear mortal, the leader is perceived, albeit unconsciously and unrealistically, as omnipotent and capable of withstanding the most fervent anger. Therefore, groups members can directly confront her without out fear of wounding her.

Disunity. The inability of the group or the leader to satisfy member expectations leads them to the first conflict stage, Disunity. Transference is prevalent. Group members view leaders with preformed ideas of how authority figures should act. Members withhold anger and avoid fighting with the leader for the same reason they are unable to confront their parents. They believe their survival, figuratively, and sometimes literally, is predicated on appeasing these powerful figures. However, no matter how well the leader leads, she can never satisfy all of the expectations of the group members. When the leader fails to meet expectations, members react with frustration, anger, and confusion. Once the group has moved beyond the initial sessions and a minimum level of cohesiveness has been established, the leader works to facilitate the open expression of anger. This is particularly true when she suspects that either the group or a member is angry with her.

At this stage, frustrations surface in the form of indirect anger. Members begin to unconsciously challenge and resist the leader's interventions. Boundary testing occurs as members question the group structure. Indirectly, they are also challenging the strength of the leader and her predictability. Furthermore, they are assessing her ability to handle conflict. These challenges are further means of establishing a safe and predictable environment.

Intermember rivalry surfaces, very tentatively and beyond the immediate recognition of group members, as power issues are contested. Members suggest actions, demonstrate leadership capabilities, and subtly vie for the leader's position. Politeness is abandoned; commitment to the group is questioned; willingness to commit to intimacy is explored. Scapegoating or subgrouping may occur. As conformity gives way to difference, divergent views are expressed.

Metaphors may surface as members covertly express dissatisfaction with the leader.[31] Individuals feel very constrained at this point in the group. During this phase, a primary responsibility of the leader is to invite and encourage the direct expression of anger. As anger is expressed, a risk by one member to openly challenge the leader may be the catalyst that sends the group into Stage Four.

The group's development ceases without the expression and resolution of conflict. Unfortunately, many groups never progress beyond this point. Often it is due to the leader's lack of understanding about the importance of conflict in developing groups and/or her own discomfort with it. Group members, too, may avoid the anger because of their own unease with it.

Groups at this stage of development are psychologically immature. Most groups are capable of evolving beyond this stage; however, many fall short of achieving their full potential and remain stuck in the initial stages.

Conflict/Confrontation. Anarchy characterizes this stage. Social convention is abandoned as the group attempts to dethrone the leader. Attacks on the leadership are over issues of power and control and dependence and independence, and are often based on transference. Group members confront their capacity for governance while attacking the leader's early group role as parent and protector. The anger and frustration directed at the leader are not personal. Although they may be experienced as personal, these feelings are directed at her role in the group. Leaders often miss this point and retreat from engagement when they misinterpret the attack as personal. Bear in mind, members are frustrated with the leader's role and her inability to meet their leadership expectations within the group.

The norm for the expression of anger and conflict resolution is initiated by the leader in this stage, then forged during confrontation with the members. This norming process is described in the next chapter.

If the group leader's only experiences with conflict have been negative and she avoids conflict in the group, then she must work through this reluctance with her supervisor. Bear in mind that the group's boundaries are shaped by the leader's own psychological limitations. The more the leader endeavors to transform her personal barriers, the more opportunities the group has for movement and growth. Most people have difficulty dealing with anger; the group leader is no exception. During this stage, group members must come to terms with their own capacity for leadership by confronting the leader's position in the group. This confrontation must occur before the group takes control of its own destiny.

The turning point in many groups rests on the leader's ability to confront members on their dependencies. The leader must then relinquish the leadership position so group members can experiment with assuming that role and learn "how to exercise mature power in a group setting." Ideally, ". . . the mature group is composed entirely of leaders."[32]

Timing is critical. The leader must not relinquish power before the group is ready. She must assert her position, and the group must engage and overpower her to earn the rights to leadership. The empowerment of members comes from their ability to conquer this omnipotent leader. Confronting, engaging, and vanquishing the leader all contribute to the transfer and investiture of leadership authority in the members. Leadership must be earned!

Placement at the arc's vertex underscores the criticalness of this stage. Overall, this is the stillpoint in the arc, the transition period that signifies the shift in power from the leader to the members. By this stage, the group is totally constrained by their frustrations with the group and the leader. They face two choices: open confrontation with the leader or retreat into regressive solutions.

The road to freedom is through the fire, not around it. Although flight into a regressive solution (chapter 9) remains a viable option, if the leader is effective, the group attacks. Her ability to invite, facilitate, and withstand the encounter establishes the crucial norm for conflict resolution and propels the group back up the arc.

Disharmony. Member disagreements surface fully once the leader is overthrown and a norm for the expression of conflict and conflict resolution is established. Having witnessed and survived conflict, members are now prepared and free to resolve their differences, differences that were suppressed during the descent for the sake of unity. Now, as members experience their independence from the leader, they can assert themselves and individuate from one another. The expression of anger tests and confirms the strength of their relationships with one another. Sharing feelings of anger presents less of a personal risk than sharing more intimate expressions of affection. Anger is a feeling, but it keeps members at a safe distance and does not invite the deeper connections that come with expression of more vulnerable feelings. If members are perceived as being able to tolerate and withstand one another's anger, and not leave the group, then the expression of more intimate feelings is possible. Differentiation among members and respect for difference allows the collective to regain a degree of freedom previously lost.

All members must fully join the group at this stage for group development to continue. In order for members to pursue intimacy safely, affiliation and dependency issues must be resolved. Full commitment to the group is not possible without resolution. Members who withhold and are self-protective cannot wholly participate in the group.

In nontraining groups, members who found the group too unsafe or stressful probably quit by this stage. There are numerous reasons why this occurs. Many individuals are not ready to confront certain issues in their lives and should never be required to do so.

However, there are cases where a marginal member has managed to be sustained by the group. In these instances, the group projects meaning onto the person's resistive behavior, based on minimal cues. These projections may or may not be true. However, they create the illusion that the resistant member is conforming to current levels of self-disclosure in the group. This projection process enables members to continue in the group with a minimum of input.

Marginal members sometimes manage to escape some of the group anxiety by being late or absent from meetings. However, at this point in the group the reluctance must be confronted. On very rare occasions, this may be accomplished outside of the group. If the member is feeling too unsafe within the group, a private conversation with the leader may be helpful. This should only be undertaken as a last resort, and should come at a time when the leader is facing the decision to retain or dismiss this member from the group. The meeting, not the details, should be reported to the group. When all members have equally invested in the group, a collective "we" emerges.

Harmony. This stage is characterized by feelings of relief and, occasionally, euphoria. High morale, respect, mutuality, and group pride mark this stage. There is a great sense of accomplishment, having successfully managed the conflict phase. It is also a quiescent period during which the group rests in preparation for more work. This is usually the shortest stage in the group's development, similar to a way station, where members stop for a brief rest to bask in their accomplishments, recoup their strength, then move on.

Members value the group. They feel connected to one another and often talk in terms of "we." A bond was established among them. Pride in the group's accomplishments is expressed. Members experience a sense of mastery and achievement as they come to fully appreciate the group experience and see benefit in it. There is a strong desire to continue the group. When describing the group, members speak of it in glowing terms.

Members interact with one another rather than with the leader. They share leadership functions and take over boundary and maintenance functions. The leader becomes a consultant to the group.

The monitoring behavior exhibited in the earlier stages is virtually nonexistent. Members trust their judgment and understand the norms that were established. Both negative and positive emotions can be equally shared.

Personal needs are expressed as the group listens objectively. Intimacy marks this stage as members begin to disclose hidden aspects of themselves. Members are seen and accepted for their differences. In juxtaposition to the superficial Unity stage, here diversity is valued over conformity. The freedom to be themselves is regained. Opportunities for intimacy exist and offer further gain for the collective. However, in the euphoria of the moment, sometimes members minimize differences to embrace the harmony. As a result, not much work gets accomplished, and if the group gets complacent, it is the leader's task to nudge them forward.

Performing. The final stage, Performing, centers around productivity. The group works in the here-and-now. Authentic relationships are possible as members share honest and direct feedback with one another. Earlier fantasies and projections are openly examined. Members are safe to explore themselves at deeper levels. The group is free to focus on members with greater objectivity because personal relationships were established. If members are willing to risk disclosure, a more real sense of self emerges, leading to greater independence and freedom.

One way this is accomplished in the performing stage is through acknowledgment and acceptance of one's dark side, that aspect of the personality kept hidden from others. In this safe environment, the opportunity to reveal oneself fully and experience the acceptance of others can be cathartic. It is remarkable to witness group members reveal some aspect of self that they have hidden from others, only to find that the disclosure has drawn people to them, not pushed them away. These group self-disclosures facilitate self-acceptance, the essence of self-actualization. Profound and lasting change can be accomplished during this stage. The change is akin to restructuring self-perception, rather than simply altering behavior.

The group is capable of extraordinary healing. Peak experiences are possible. The group flows and appears to function effortlessly as members share leadership responsibilities. Freedom that was relinquished by individuals during the descent is fully regained.

Summary

Every group is unique, comprised of individuals with differing needs. So the length of time spent in any one stage is, in part, dependent on the composition of the group members. In addition, the group leader's understanding of group dynamics and skill in recognizing transition opportunities or high leverage points can advance or impede group development.

ISSUES OF CONCERN

To understand how energy or creative tension is generated in developing groups, a second arc, comprising issues of concern, is depicted in Fig. 3.8. The issues selected represent those described in many of the theories

Issues

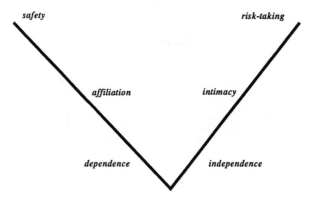

FIG. 3.8. Issues arc.

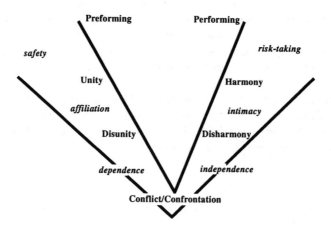

FIG. 3.9. Group and Issues arc.

reviewed earlier. Figure 3.9 depicts the issues of concern juxtaposed against the group stages. There is an approximate association between the group's ability to move to higher levels of development and the satiation of individual issues in each stage. Energy, amplified by the push to form the collective and the pull of individual issues, propels the group forward. These centrifugal and centripetal forces create the spirals of turbulence depicted in Fig. 3.10.

The age-old image of the spiral as both a symbol of change and a transformative container for collective self-organizing is explored at the end of this chapter. It is sufficient to note here that the spiral image is ubiquitous and even appears in several studies referenced in this book. Remember Gottman's research, in the previous chapter, and the spiral

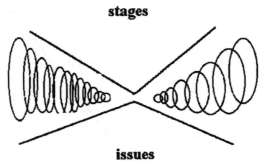

stages

issues

FIG. 3.10. Arc spirals.

that emerged from connecting the plotted points in Fig. 2.5. Likewise, Gemmill and Wynkoop use the spiral as a transformative container for collective self-organizing in their model of group change discussed in the next chapter.

In each stage of development, one of the group's tasks might be viewed as satisfying the stage-appropriate issue of concern: In the Preforming and Unity stages, it's safety; Unity and Disunity, affiliation; Disunity and Conflict/Confrontation, dependency; Conflict/Confrontation and Disharmony, independence; Disharmony and Harmony, intimacy; and Harmony and Performing, risk-taking.

Placement of these issues in relationship to the group stages is an approximation. Not all members will satisfy safety issues concurrently with group movement from the Unity to Disunity stage. Some members may not fulfill them until the Disharmony stage. Inclusion of this issue hierarchy is a reminder that members are undergoing their own intrapsychic struggles concurrent to those experienced by the group. It also illustrates a competing force against which the evolving group is organized.

The spirals depicted between the arcs in Fig. 3.10 symbolize the organizing or coevolution that occurs in groups between individual and collective needs. Small and large scales simultaneously influence one another. The back and forth movement between these scales produces the forward momentum that propels the group. In Jantsch's evolutionary spiral (Fig. 3.11), we witness the same interaction between micro and macro levels at the planetary level. Like the group spirals, Jantsch's model depicts the transformation to higher levels of development. Each twist of the spiral represents greater complexity and autonomy as evolution advances. Reiser's spiral of evolution (Fig. 3.12) provides another view. Likewise, the group model depicts this advancement. As Laszlo proposed, greater structural complexity, higher levels of organization, and increases in dynamism and autonomy result as the group progresses through the arc.

Although the spirals depicted in these models appear smooth, the movement they represent is not. "Co-evolution is full of chaotic order in which large-and-small changes mirror each other, jumping back and forth, producing an evolutionary movement that is unpredictable but completely interconnected."[33]

Placement of these issues can also be viewed as transitional points. When groups regress during times of stress, members can be seen as retreating to the closest transition point, exhibiting an appropriate issue of concern behavior. Thus, if excessive stress is encountered in moving from the Disunity stage to the Conflict/Confrontation stage, members might draw back to affiliation issues. Likewise, if the group experiences excessive stress between the Disharmony and the Harmony stage, members may retreat to independence issues.

Macroevolution

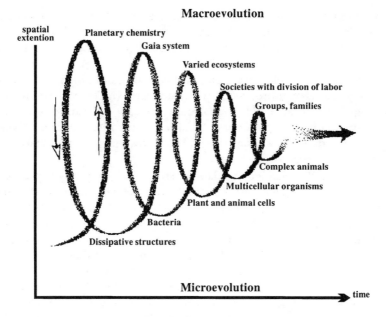

FIG. 3.11. Jantsch's evolution spiral. From *Turbulent Mirror: An Illustrated Guide to Chaos Theory and the Science of Wholeness* by J. Briggs and F. D. Peat (1989). Copyright © 1989 by J. Briggs and F. D. Peat. Reprinted by permission of HarperCollins.

Group members are affected differently by each issue and must resolve them for the group to succeed. The grouping process is the milieu in which these issues must be resolved. Each member must find a way to satisfy them within the emerging group.

Although each issue is associated with a particular stage, all of them are present and recycled through every stage. However, during the descent, safety, affiliation, and dependency issues occupy the group's energy. The complementary issues, independence, intimacy, and risk-taking, although present, remain buried. In the ascent, the reverse is true. Independence, intimacy, and risk-taking issues occupy center stage, whereas safety, affiliation, and dependency issues are hidden. Of course, certain issues may still predominate for individual members. In the ascent, one or two members may still be resolving safety or dependency issues while the group focus has shifted toward stage-appropriate issues.

In each stage, one issue is illuminated over the others. During that period, circumstances and timing are such that the potential for resolution of that issue is maximized. Throughout the group's history, these issues of concern are revisited again and again. As the group gains new awareness about its own processes at each succeeding level of development, new meaning is given to these issues.

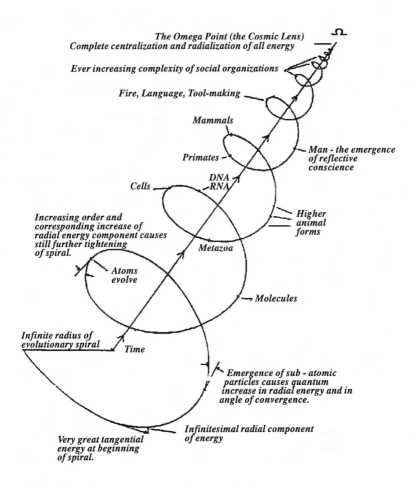

FIG. 3.12. Reiser's evolution spiral. From *Pain and Possibility* by P. Rico (1991). Copyright © 1991 by G. L. Rico. Reprinted by permission of Jeremy P. Tarcher, Inc., a division of The Putnam Publishing Group.

During the descent, group members are influenced by issues of safety, affiliation, and dependency on the leader. Movement is inward, centripetal; the descent represents contraction. Members conform in an effort to belong, to be included. They are cautious, anxious, and seek direction from the leader. Members want security and predictability, and externalize these needs to the group. Initially, it is the leader's responsibility to create a safe and predictable environment. If a safe enough environment can be created, members are eventually able to internalize or introject these security feelings. The thrust for collective organization is constrained by the pull for satiation of the primary issues of concern. Individual survival is predominant over collective survival during the initial stages. Sufficient numbers of members must resolve (phase lock) the primary issue at each stage for the group to progress.

Affiliation issues are central in developing groups. Initially these issues are expressed by members as they search for a connection or role in the group. Their sense of safety is increased through connection and belonging. In the ascent, intimacy issues provide opportunities for deeper levels of relationship. Both affiliation and intimacy requirements are often expressed as inclusion and exclusion themes. Throughout the group's life span, members move through escalating levels of relationship based on trust. The development of each new level of trust occurs as members disclose more and more about themselves. Membership in each new level requires that all members disclose a similar amount. This varies from stage to stage. Some members are hesitant and hold back. They can be carried along, but only until Stage 5, when all members must share a similar level of self-disclosure.

Each succeeding relationship level requires more sharing on the part of the group members. The group may proceed rapidly at first, and then slow to a snail's pace as more and more is required of the membership. The intriguing aspect of this spiraling process is that group energy is continually focused on inclusion. Groups expend great energy working to include all members as each new level of familiarity is obtained. Even in the face of the most recalcitrant members, the group works diligently to include them. Only after exhausting effort that has met with no success will the group seek to exclude one of its members.

The leader's role is to monitor the progress of the group and the inclusion of its members. Attending to silent members, regulating the speed of self-disclosures, attending to feelings are methods of facilitating inclusion. As the group progresses, the leader must continually ask himself, "Is everyone included?"

Beginning the ascent, most members have resolved primary issues. Secondary issues of concern, or more appropriately, opportunity issues, now emerge. Opportunities for independence, intimacy, and risk-taking are available to be experienced in the group. Much of the therapeutic value of groups is found on the right side of the arc. It is gained through self-assertion: emphasizing difference and independence, fostering intimacy, and risking-taking through self-revelation.

Individual movement during the ascent is centrifugal—it pushes outward toward separation. Group survival is now predominant, and the inward pull is to remain safe in the secure group environs. For growth, members must take risks and move away from the group sanctuary.

Unlike the primary issues, secondary issues do not have to be satisfied by all members. However, a critical number of members must assert themselves at each stage for the group to continue its upward path.

The symmetry of the individual-issues arc is self-evident. Inward movement during the descent is countered by the outward thrust of the ascent. Safety and risk-taking, affiliation and intimacy, dependence and independence are complementary issues. Each is part and whole of a larger totality. They mirror one another, albeit left arc themes are reflected outward, requiring inward resolution, and right arc themes are reflected inward, requiring outward resolution.

Arguments can be made suggesting alternatives or perhaps a different ordering of the needs. However, what is of primary importance is that the issues arc illustrates the dialectic inherent in groups between individual and collective needs.

The issues arc draws from one of Maslow's ideas. Issues depicted on the left side of the arc are primary issues and must be resolved before individuals can obtain benefit from higher level or secondary issues depicted on the right side. Unlike Maslow's hierarchy, the primary issues do not have to be satisfied in order. However, as discussed, they must be resolved by the Disharmony stage.

Secondary issues, indexed on the right side of the arc, reflect opportunities for growth and change. All group members do not have to fulfill them for successful completion of the group; however, the overall group experience is heightened significantly if they do. One of the reasons people leave groups with varying degrees of experience relates to how successful they were at satisfying secondary issues.

FIRST CAUSE

An important group dimension seldom discussed is the transpersonal or spiritual component. It is difficult to explain. I will try not to complicate it. Mostly it is a shared experience that cannot be verified by traditional methods. Although very few groups ever directly access this sphere, it is important to recognize its existence. Just as individuals are capable of transcendent experiences, so are groups. The spiritual is not to be confused with religion. Although aspects of each may overlap, the spiritual embraces the mysteries of life, whereas religion interprets and organizes them. One manifestation of the transpersonal is the group mind explored fully in chapter 10.

The first stage, Preforming, is included to acknowledge the existence of this spiritual realm. Young's notion of first cause, or teleology, is relevant here. The group originates from an Indivisible Unity and exists as yet-unrealized potential. The potential manifests as purposive action and propels the group to the arc. The arc contains the learning stages through which this potential must pass in order to become actualized.

Let me clarify this idea by describing the artist Jackson Pollack's work, characterized as action painting. Oftentimes he would stand in the middle of a large canvas, placed on the floor of his studio, and fling paint on it. Aesthetics aside, Pollack's stroke parallels the idea of first cause. As artist, he imparted his intention into the flinging motion. As the paint left the brush, a brief moment between cause and effect occurred "that was out of control. Like the gap in a sparkplug, this moment is what Aristotle once proposed was the potentia."[34] *Potentia*, for Aristotle, was the brief moment in which the spontaneous could occur. Shlain says Pollack understood "that this gap . . . is the crack in the cosmos through which all things and images enter the extant world of manifestation."[35] This is exactly the perspective echoed by the poet Leonard Cohen when he sings, "There is a crack in everything, that's how the light gets in."[36]

This description captures the idea of first cause. In the Preforming stage, the group is part of an initial unity, albeit unconscious, with the Indivisible Whole. The original spark that propels the group forward, manifest as intention in the group, contains the realization of this connection to the Whole. Through the arc, the group exists as potentia. No two groups form, develop, or traverse the arc similarly. However, the spiritual potential of each group lies in its ability for members to merge in a generative collective that recognizes their unity with and connection to what many describe as a collective or universal wisdom.

They progress through the various stages where members struggle to gain awareness of the original connection. Throughout the arc, members are transformed at each stage, learning as they develop, gaining knowledge about the group and themselves. The V shape of the arc implies an indirect route to the end goal, but learning is a process that takes time. Successful group members gain full awareness of their relationship to the Whole by the Performing stage. This spiritual process indicates a vision of groups, inherent in this model, as a journey that, when complete, returns to where it began. The arc symbolizes the group's journey. Metaphorically, the arc represents Campbell's idea of the heroic journey. In this case, the group traverses the arc to discover its essence or realize its potential and returns to where it began, with new and fuller awareness.

There are other ways to consider these transpersonal ideas. One metaphor for the group's spiritual journey is the Hindu idea of reincarnation. Simply put, the soul exists as pure light or energy, in full awareness of its connection to the Indivisible Whole, waiting to be reborn. Once reborn in a physical form, the soul loses this awareness. Over a lifetime, the soul seeks to regain full awareness of its original connection to the Whole.

Another way to picture this spiritual journey is to return to the image of the holographic universe. In it, group members comprise the Whole and the Whole is reflected in each member, albeit individual reflections of the Whole are very weak. However, when groups of individuals merge together or entrain with others, the light becomes brighter and the Whole vision is made clearer. Groups provide one means by which we can magnify our relationship to the Whole.

Whether or not you agree with these ideas shouldn't lessen your desire, as a leader, to encourage group members to engage their curiosity about the group's purpose. Why us? Why here? Why now? What am I to learn about myself? At the very least, each question stimulates attention to the group-as-a-whole and captures the imagination of many group members. Initially, combining these questions provides a larger vision that holds members together through the tenuous forming stages.

THE SPIRAL

The image of the spiral (Fig. 3.13) emerges as a central symbol in my efforts to integrate chaos theory with our understanding of small groups. The spiral is ubiquitous. It is found at all levels in nature: from the turbulent winds in the stratosphere to the hurricanes, cyclones, and tornadoes that churn about the earth to the small eddies that form in streams and brooks. It is the vehicle that shapes and transforms our innermost forces into creative energy.[37] This figure also appears in the works of Gottman and of Gemmill and Wynkoop. In chapter 2, the spiral frames the group developmental model illustrated there. It is an image of movement, possibility, and transformation.

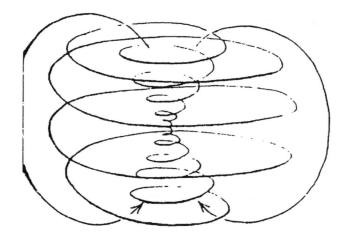

FIG. 3.13. From *Pain and Possibility* by G. L. Rico (1991). Copyright © 1991 by G. L. Rico. Reprinted by permission of Jeremy P. Tarcher, Inc., a division of The Putnam Publishing Group.

The movement, like that of the strange attractor, is dynamic: up and down, in and out, contracting and expanding. It depicts the constant folding and unfolding of order and chaos, stability and change. The spiral signifies change, and through its image we see the relationship of all things: the connection between opposites, the dialectic created between the positive and negative, each so necessary and vital for development. The spiral functions as a powerful metaphor for growth.[38]

The spiral signifies possibility. The tension present between opposites pushes and pulls and creates emergent opportunities. Movement through the spiral is nonlinear, discontinuous, disconcerting, and disorderly. It shakes up, recreates, and reorganizes in unpredictable fashion.

The spiral acts as a container. It helps us tolerate ambiguity, frustration, and paradox because in its image, as we stand outside it, we recognize that its unrelenting turbulence is harnessed and constrained by the emergent pattern of the spiral. And, as Rico realizes too, from the bigger picture we are able to see that change and stability are both necessary parts of life.

Rico sees a double spiral in the spherical vortex and, "like a hurricane or tornado, it is both implosive and explosive."[39] For her, the implosive cycle reflects a disintegration, a movement inward signaling a "dark night of the soul" or "existential crisis." But she also recognizes in the collapse the emergent possibility of a breakthrough to new insight or an epiphany leading to a life change. The explosive cycle reflects the simultaneous experience of breaking apart and coming together. Rico eloquently captures this notion of synthesis ". . . taken together we realize that beginnings are endings, endings are beginnings, which is to say all possibility is present in a given moment."[40]

At the threshold between the implosive and explosive cycles is the *stillpoint* or, as Rico calls it, the "gateway to possibility." Here the shift originates with the stillpoint, or liminal period, where chaos enfolds into order and order enfolds into chaos. Briggs and Peat reintroduce us to the notion of nuance that can be found in the "personal subtleties of tone and meaning for which we have no words."[41] They assert that "in experiencing nuance we enter the borderline between order and chaos, and the nuance lies in our sense of the wholeness and inseparability of all experiences."[42] This is the stillpoint. As we more closely examine the transformational process of small groups and the role played by the group facilitator, we return to these very useful ideas of nuance and stillpoint.

The significance of the spiral comes from both its historical and mythical meanings. Further insight into this symbol can be found in Purce's narrative, *The Mystic Spiral: Journey of the Soul.*

Chaos and Self-Organization
in Groups

Group development is characterized by periods of relative calm punctuated by intervals of chaotic activity. This periodicity is essential for growth and reorganization, for without undergoing periodic upheaval, groups cannot evolve. Whereas some change, described earlier as first-order, is incremental and relatively predictable, the kind of change that introduces novelty into groups is discontinuous. This second-order change describes the movement from one stage to another in developing groups. Disorder precedes second-order change, and its outcome cannot, with any certainty, be predicted from previous conditions. Understanding how groups undergo this metamorphosis is essential for effective group leadership, because attempts to control and limit it lead to regressive and potentially destructive solutions. As we learned in chapter 1, social systems are self-organizing.

GROUP CHANGE

At the heart of what I consider the most valuable element that we can take from chaos theory and transfer to our work with groups is the insight relative to (a) change in systems far from equilibrium and (b) self-organization. In chapter 2, different theorists, utilizing principles from chaos theory, described the change processes in the psyche, families, and society. In this chapter, many of these ideas are further clarified and used to describe how change occurs in small groups. The critical factor in this change process, phase-locking, is closely examined. Koestler's notion of biosociation is introduced to further vivify and elucidate our understanding of discontinuous change. Finally, Gemmill and Wynkoop's ideas about first- and second-order change in groups are presented.

Combined, the ideas presented in this chapter offer an important vision of group development: At the root of successful group leadership lies the ability to recognize transition points and to facilitate change. Let us begin this discussion of phase-locking with a brief overview of how groups progress from one stage to another, in this case from Unity to Disunity. Remember that the process of change is equivalent across all levels or scales of group. Therefore, the principles of change at one level can be applied to all levels (e.g., intrastage change, intrasession change, etc.).

STAGE TRANSITION

In the Unity stage, individuals come together to form a group. Members are asked to invest themselves in a collective process that offers little instant gratification. The central theme each member must resolve is safety. Negotiations over goals, rules, and boundaries begin immediately as members go through the initial sorting process to establish and secure their working environment. In the forming stages (left side of the arc), when anxiety is high, groups retreat to the latent level and conduct many of their negotiations outside of their conscious awareness. A discussion of the latent and manifest levels of group communication is on pages 92 and 93. It is important to make this distinction between these two levels because the changes that groups undergo normally occur outside the conscious awareness of the members. This is particularly true throughout the initial stages. During the ascent, when members assume responsibility for the group, their ability to recognize latent level nuances that prescient change dramatically increases. Leaders and group members who become skilled at attending to the subtle nuances found in latent level metaphors can successfully guide groups through turbulent change periods.

Much of the overt discussion in the Unity stage centers on past experiences. In this way, members are provided initial data about one another, but more importantly, members can avoid the angst of the present. Occasionally, the group wades into the present but retreats when anxiety deepens. This movement back and forth, between the safety of the past and the angst of the present, characterizes this stage. The spirals (Fig. 4.1) depict the group's movement. The chaotic lines at the top indicate the initial lack of accord among members. Members enter the group with various security feelings. Some with previous group experience understand and better tolerate the ambiguity inherent in this initial stage. Others are less secure, both with themselves and with others.

Early stage development

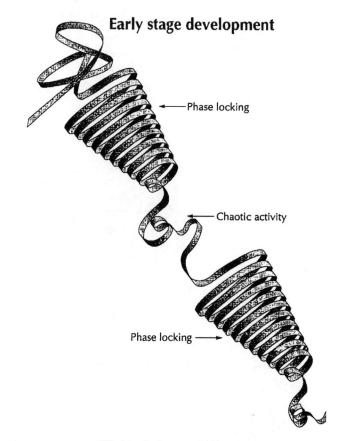

FIG. 4.1. Early stage development.

As members interact, the unknown is made known. Hypothetical fears, fantasies, and projections are slowly replaced with objective data. Members learn about one another. They self-disclose, investing in each other and in the group. As investment and group predictability increase, the group orchestrates its dance. The wider cycles at the top of the spiral in Fig. 4.1 depict this beginning. As group norms and rules of behavior emerge, the group enters the spiral's vortex.

As safety increases, members phase-lock, depicted by the tightening spiral in Fig. 4.1. When a critical mass of phase-locked members is reached, the group leaps to a new level of organization, in this case Stage 3 (Fig. 4.2). Then the process begins anew. Later stage development is depicted in Fig. 4.3 by tighter spirals that reflect a more cohesive group.

Keep in mind that not all members must phase-lock for the group to evolve, only a critical number. Reluctant members can be carried along until Stage 5. Critical mass varies and likely increases from stage to stage.

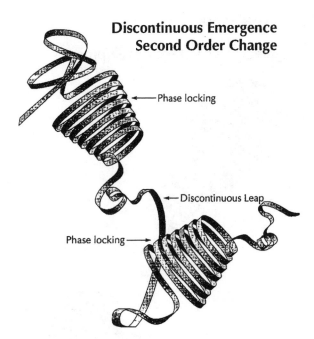

**Discontinuous Emergence
Second Order Change**

←— Phase locking

←— Discontinuous Leap

Phase locking —→

FIG. 4.2. Discontinuous change in groups.

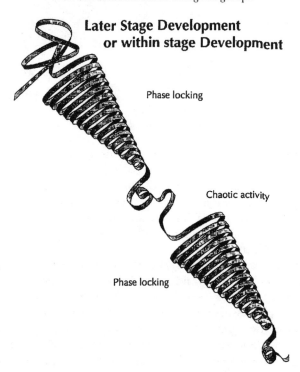

**Later Stage Development
or within stage Development**

Phase locking

Chaotic activity

Phase locking

FIG. 4.3. Later stage development.

Initially, a few members may be able to carry the group into Stage 3. At each successive stage, more self-disclosure, intimacy, and commitment is demanded from members. As such, a larger critical number is probably needed to propel the group forward, until Stage 5, when all members must phase-lock for the group to continue.

PHASE-LOCKING

Phase-locking, as discussed, refers to a kind of harmonic convergence wherein frequencies from individual actors are entrained, leading to a coordination of actions or collective behavior. The slime-mold and spatial organization of the B–Z reaction detailed earlier were examples of these phenomena.

Phase-locking offers a very novel idea about how individuals in groups might evolve toward collective action. Let us briefly examine a research study that might clarify our understanding of phase-locking. It was cited in Briggs and Peat's terrific book *Turbulent Mirror: An Illustrated Guide to Chaos Theory and the Science of Wholeness*,[1] and was conducted by Michael Guevara, Leon Glass, and Alvin Shrier at Montreal's McGill University. Their research involved two stages. In the first stage, they extracted cells from the heart of a chick embryo and placed them randomly in a solution. The cells continued to beat erratically. In a couple of days, the cells had phase-locked, producing a collective oscillation.

In the second stage, the experimenters introduced a series of pulses into the cell aggregate with an electric probe. The heart cells phase-locked on the incoming signals and produced a steady pulse. As the experimenters modified the electrical frequency, they were able to perturb the cells into period doubling and then chaos. Briggs and Peat concluded that human hearts operate similarly. Individual heart cells phase-lock, producing a heart beat. Natural pacemakers, such as nerve nodes, produce electrical signals that drive and organize the cells. At rest, the stability of a phase-locked heart is advantageous. Frequency change is needed for sudden bursts of activity; here the nerve nodes or pacemakers function perfectly. If heart cells can synchronize their rhythms, what about the brain?

In a study conducted at the Salk Institute, Ehlers found a strong similarity between EEG results and behavior. She asserted, "Maybe what brings the whole system together from behavior to the EEG down to enzymes" is a frequency organizer. She cited Mandell's idea that "frequency is a basic language, a global property. . . . A description of consciousness may be a description of the variability of mental stages and the organization of those states in time."[2] As we see from Freeman's work, the brain is self-organizing, and perhaps it is brain waves or frequencies that enable us to entrain and organize our thoughts and behavior cooperatively with others to form a collective mind.

Adey saw the brain's language as the synchronized "whispering together" of millions of neurons. The brain communicates in these wavelengths, not in impulses. A fascinating aspect of one of his studies confirmed for him that individual brain cells are sensitive to surrounding electromagnetic waves. He found that if a weak electromagnetic wave was applied to the head, "neurons would synchronize their firing to the surrounding rhythm."[3] In his work with monkeys, Adey was able to influence the animal's alpha wave production by placing it in an radiomagnetic field. He suspected these fields may influence our circadian rhythms. Then what are the implications from all the weak electromagnetic fields in which we live? The computer screen that I am staring at now, telephone lines, cellular phones, radar installations, and microwave products all produce electromagnetic waves sufficient to alter the brain. Could our moods be influenced by these waves?

The brain's capacity to produce and be receptive to electromagnetic waves, combined with its ability to self-organize, may provide some of the clues necessary to explain the emergence of social order. However, regardless of how it happens, there is substantial evidence that it does occur.

Let us return to Briggs and Peat for their concluding observations. As objects (quantum matter, cells, organisms, chemicals, individuals, etc.) come together and phase-lock, there emerge certain stable and definite properties of the collective. In other words, "The collective can no longer be described by a linear combination of different states."[4] It now becomes described by its emergent characteristics (e.g., individual amoebas of the slime-mold become known collectively as the pseudoplasmodia). These ideas coincide with Laszlo's description of social systems as having functions and attributes separate from the individuals who comprise them.

Briggs and Peat cited a couple of common examples of phase-locking in living systems. Isolated from any changes in light, meals, or clocks, our bodies operate on a 25-hour cycle. Once in light again, our biological clock is phase-locked into the 24-hour day. However, a long airplane trip causes our bodies to get knocked out of phase, and we experience jet lag. Women who live closely together phase-lock when their menstrual cycles become synchronized. Briggs and Peat offered a fascinating observation about phase-locking developed by Bohm.

Bohm, combining ideas from the physicist Lorenz with Einstein's relativity, asserted that neither time or space are absolute. In fact, he argued, both are generated in "material frames" as a result of "the phase-locking of matter within that frame." Therefore, Bohm claimed, "Time is a measure of the amount of process that takes place, the ticks of the frame's internal clock." So if clocks are not synchronized, it is because their material frames are out of phase. Briggs and Peat speculated that the phase-locking of material frames may take place with respect to individuals and cultures, which may explain why different people and different societies operate with different "senses of time."

Let me give a brief example of relative time. Bill Russell, who played professional basketball for the Boston Celtics, described how the performance of his team was maximized. "Every so often a Celtic game would heat up so that it became more than a physical or mental game . . . the feeling is difficult to describe . . . my play would rise to a new level . . . all sorts of odd things happened . . . the game would move so fast . . . but it was as if we were playing in slow motion." What Russell describes is a peak and transpersonal experience, a moment when the team was phase-locked and created a "slower sense of time."[5] Sports are replete with such examples as Russell's.

Chaos theory switches from linear or chronos time to nonlinear or kairos time. Remember, chaos is not concerned with the linear dimensions of life. As Ainslie correctly asserted, chaos theory is concerned with the immediate, the now, "It thrives in the moment." Nonlinear or existential time is intimately connected with the spiritual and is further explored in chapter 10, Generative Groups. One other source for further exploration of constructed time is Csikszentmihalyi's book *Flow: The Psychology of Optimal Experience*. Now, let us venture out onto the proverbial limb and speculate how phase-locking might occur in small groups.

PHASE-LOCKING IN GROUPS

What does it mean to phase-lock in groups? At the manifest level, phase-locking can be pictured as commitment. Commitment is evidenced by a member's willingness to invest himself or herself in the group. Before the group has established common expectations, commitment takes many shapes. Manifest behaviors range from self-disclosure, to showing up on time and being in attendance, to expressive nonverbal behavior. During the descent, commitment at the manifest level is often subjectively interpreted by other members. Hence, perception commonly substitutes for reality. During the ascent, objective verification of commitment becomes the norm. More importantly, overt behavior is but a manifestation of a connection at the latent level that synchronizes, or phase-locks, members. In chapter 10, the works of Bohm and de Chardin are used to illuminate the relationships between the exterior and interior reality.

CONSTRUCTIVE INTERFERENCE

Suppose we consider each group as part of a holographic universe. The group would be divided into an implicate and an explicate order. Were you able to observe a group through Bohm's lens, members in the implicate order would replicate the whirls of color found in the bubble chamber: pure energy, no distinctions among them. At the explicate level, we would witness what we call objective reality: Members sitting on chairs, in a circle,

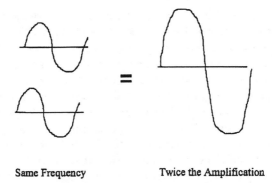

Same Frequency   Twice the Amplification

FIG. 4.4. Constructive interference.

facing one another. In the bubble-chamber, we exist as one form of energy, a lightwave. When lightwaves interact, interference patterns result (Fig. 4.4). Lightwaves that reinforce one another are said to constructively interfere; their frequencies are synchronized and their energy is amplified. Constructive interference is phase-locking.

Senge's illustrations, from his book *The Fifth Discipline*, provide another view from which to consider these ideas. As the group develops, members begin aligning (Fig. 4.5). When group members "align" (constructive interference) and their energies harmonize, a shared vision results (Fig. 4.6). Alignment amplifies individual energy or light into the coherent power of the laser, producing the collective vision.

DESTRUCTIVE INTERFERENCE

When lightwaves cancel one another, or destructively interfere, they remain out of phase (Fig. 4.7). No amplification occurs. When members are unable to harmonize, their collective energy remains dissipated (Fig. 4.8). Senge

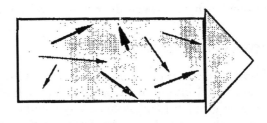

FIG. 4.5. Aligning group members. From *The Fifth Discipline* by P. M. Senge (1990). Copyright © 1990 by P. M. Senge. Reprinted by permission of **Doubleday**, a division of Bantam Doubleday Dell Publishing Group.

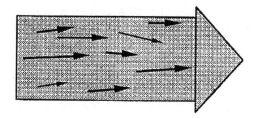

FIG. 4.6. Aligned group members. From *The Fifth Discipline* by P. M. Senge (1990). Copyright © 1990 by P. M. Senge. Reprinted by permission of Doubleday, a division of Bantam Doubleday Dell Publishing Group.

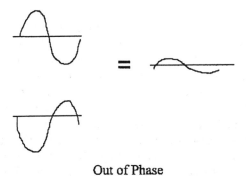

Out of Phase

FIG. 4.7. Destructive interference.

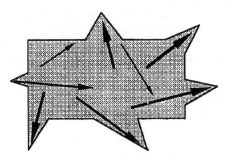

FIG. 4.8. Misaligned group members. From *The Fifth Discipline* by P. M. Senge (1990). Copyright © 1990 by P. M. Senge. Reprinted by permission of Doubleday, a division of Bantam Doubleday Dell Publishing Group.

compares this incoherent energy to the "scattered light of a light bulb." Most destructive interference is created by fear. Gottman describes fear reactions as causing *diffuse physiological arousal* (DPA). The body experiences increased sympathetic and parasympathetic nervous system activity, increased endocrine activity, and decreased immune response. All of these interfere with the organism's ability to process information. During the descent, anxiety is common and DPA arousal prevalent. Remember, initially, individual survival takes precedence over the collective. Members are cautious and are quick to retreat when distressed. Skillful leadership is needed to contain the anxiety while the group secures itself. When groups are unable to phase-lock, they remain stuck in the upper reaches of the spiral (Fig. 4.1). Group communications are recycled, roles exchanged, and metaphors created as members work to resolve their dilemma.

MUSICAL METAPHOR

Musically speaking, let us consider a philharmonic whose members are tuning their instruments prior to performance. What we initially hear is a cacophony of sounds. When the concert begins, the conductor synchronizes the separate instruments into a symphony. The collective sound is amplified into often extraordinary music (e.g., Pachelbel's "Canon in D"). Were the orchestra members unable to synchronize their instruments, sound would remain as noise and amplification would not occur.

In our musical metaphor, we might even conceive of group stages as a musical scale with each stage requiring a different phase-locking frequency (e.g., Brubeck's "Take Five" in 5/4 time, or Shostakovich's 6th, "Leningrad ...."). As group members progress through the stages or scales, the frequency increases or a higher level of commitment is required.

Phase-locking or constructive interference increases amplification, perturbing the group toward transformation. When a critical number of members have phase-locked, the group jumps to a new level of complexity where the organizing process begins anew. Leaps to new levels of organization are discontinuous. They are not linear. Discontinuous change is a peculiar notion. In *The Act of Creation*, Koestler (1990) best illustrates this idea in describing the creative process.

KOESTLER'S BIOSOCIATION

Koestler introduces the concept of biosociation, which refers to the interface between two very different frames of reference. He cites the discovery of Archimedes' Principle as an example of biosociation.[6] The story follows that Archimedes was perplexed by the problem of having to measure the amount of gold in the king's crown without melting it down. Apparently a flash of insight occurred when he stepped into his bathtub one day. He suddenly realized that by putting the king's crown into the water, he could ascertain the amount of gold in it by measuring how much water it displaced. Koestler believes the solution occurred as a result of the juxtaposition of the two different frames of reference, the measurement problem and Archimedes' bath.[7]

Figure 4.9 illustrates Koestler's diagram of biosociation. In it, he depicts an artist's search for resolution to a difficult problem. The starting point is S. You can see the mind's initial attempt at resolution as its cycles through its habitual thinking patterns. However, the solution (T) lies in another reference frame, depicted in Fig. 4.10 by the vertical plane. As the artist becomes intensely interested in the problem, her mind is pushed away from the initial limit cycle and toward disequilibrium, represented by the line connecting to the letter L. L represents the link or bifurcation point between two different frames of reference. The link represents the factor that gets amplified, perturbing the mental system toward the new solution or new mental map. The leap or jump from one plane of reference to

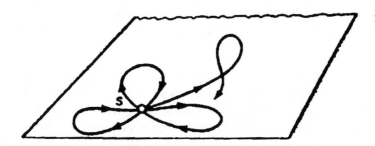

FIG. 4.9. Mind searching for solution (S). From *Turbulent Mirror: An Illustrated Guide to Chaos Theory and the Science of Wholeness* by J. Briggs and F. D. Peat (1989). Copyright © 1989 by J. Briggs and F. D. Peat. Reprinted by permission of HarperCollins.

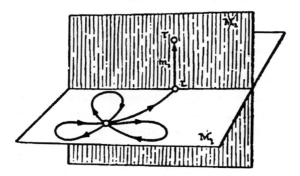

FIG. 4.10. Mind finds solution (T) in matrix 2 (M2). From *Turbulent Mirror: An Illustrated Guide to Chaos Theory and the Science of Wholeness* by J. Briggs and F. D. Peat (1989). Copyright © 1989 by J. Briggs and F. D. Peat. Reprinted by permission of HarperCollins.

another perfectly illustrates the idea of second-order, or discontinuous, change.

Briggs and Peat translated Koestler's diagrams into chaos terms. In Fig. 4.9, the initial search pattern represents limit cycles moving around a fixed point, *S*. As frustration perturbs the mental system, it increases its erratic behavior, pushing it beyond the limit cycles toward disequilibrium. In that far-from-equilibrium state, a bifurcation point, *L*, is reached (Fig. 4.10). There, a small perturbation or piece of information is amplified (e.g., water displaced as Archimedes stepped into the bath), leading thought to a new plane for possible insight into the problem. In the example in Fig. 4.10, the solution (*T*) happens to reside in that plane.

Briggs and Peat described how Gruber has taken Koestler's idea further by suggesting that the mind searches many different frames of reference when seeking a novel solution. The movement back and forth between new reference planes and the old one destabilizes habitual mental patterns, moving the mental system far from equilibrium. At that point, feedback from several new planes may phase-lock, self-organize, and produce a novel solution.

One factor that seems to increase the artist's ability to maximize this mental searching process is sensitivity to nuance, which is defined as "a shade of meaning, a complex of feeling, or subtlety of perception for which the mind has no words or mental categories."[8] "In the presence of nuance the creator undergoes what might be called an *acute nonlinear reaction* [italics added]."[9] In the next chapter, we take up this idea of nuance, in conjunction with the group leader's ability to effectively navigate in groups undergoing second-order change.

DISCONTINUOUS CHANGE IN GROUPS

The illustration in Fig. 4.10 depicts the intersection of two frames or matrices. Each matrix refers to a frame of reference or, in Bohm's words, a "material frame." The illustration is a good metaphor to use in depicting how groups self-organize and move between stages. Let us consider the material frames shown in Fig. 4.11. In this illustration, the group-as-a-whole is currently organizing in Stage 2. Suppose there are eight members in the group, each represented by a separate limit cycle. Safety is the issue of concern around which members are organizing. You can refer to the illustration in Fig. 4.8 and also envision the members as a nonaligned interference pattern.

Figure 4.12 depicts phase-locking (constructive interference) as two, then three, and finally four members phase-lock. Notice how the united limit cycle is amplified (Fig. 4.13). We assume here that four members, in

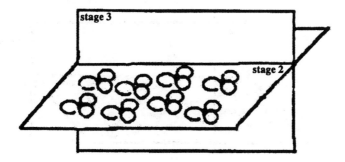

FIG. 4.11. Group organizing in Stage 2.

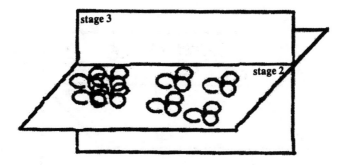

FIG. 4.12. Four member phase-lock in Stage 2.

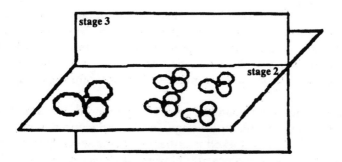

FIG. 4.13. Amplification of phase-locked members.

this stage, represent a critical mass, although that number always differs from group to group depending on relative circumstances. Once a critical mass is reached (Fig. 4.14), the group-as-a-whole leaps to another frame, in this case Stage 3, and the organizing begins anew (Fig. 4.15). Notice how phase-locked members disunite, in this new frame, returning to their own limit cycles. Whereas all members are carried along to this new frame,

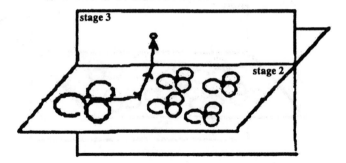

FIG. 4.14. Group leaps to Stage 3.

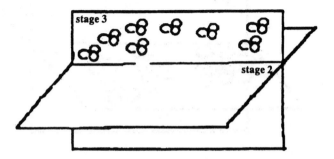

FIG. 4.15. Group organizing in Stage 3.

some have not yet resolved safety issues. Resolution of both safety and affiliation can occur in the new material frame. Remember, each new stage subsumes the previous stage and adds something to it. Each subsequent stage requires more members to conjoin as a critical mass until Stage 5, when all members must phase-lock in order for the group to jump to Stage 6.

Phase-locking at the manifest level is experienced as commitment to the group. Commitment, at this stage, can be experienced as either hanging-on (for a nonphase-locked member) or being securely in the middle (for one who has phase-locked). Movement depicted by the individual limit cycles signifies the search for commonality, security, and predictability. Overt agreement might signal phase-locking, but it does not ensure it. Keep in mind that the experience of being phase-locked remains outside of most members' awareness during the descent. Members occasionally might feel it and sense its power, but usually not fully until the Harmony stage. The magical feeling Bill Russell described was his experience of phase-locking with other team members. Groups that are unable to reach a critical mass cycle endlessly in their material frame, searching for resolutions. Eventually they stagnate into a limit cycle, and possibly a regressive solution.

The change process that groups undergo between stages mirrors within-stage development. Movement through the subphases of stages occurs in an identical manner; the scale of analysis has simply been reduced. We have moved closer to the shoreline, but the pattern remains the same.

Gemmill and Wynkoop described in-stage movement in a recent article on small-group transformation.[10] It is one example of how groups work, within stages, to resolve issues of concern. Their ideas are similar to the ones just presented, but they utilize a different model. Although their primary discussion centers on creating a model for second-order change in small groups, they hint at the importance of chaos theory for transforming our understanding of group dynamics. Let us begin by defining first- and second-order change.

FIRST-ORDER AND SECOND-ORDER CHANGE

First-order, or linear, change is gradual, sequential, predictable, moderate, and incremental. However, it constrains both inner-stage development and transition between stages. Although gradual change can have a beneficial effect, the full therapeutic potential of groups is not realized unless second-order changes occur throughout the evolution of the group.

Second-order, or nonlinear, change is turbulent and chaotic, and results in a discontinuous transformation that could not have been described or predicted from observation of initial circumstances. The transformation of the slime-mold from a unicellular amoeba to the multicelled pseudoplasmodia or of the caterpillar to the butterfly are classic examples of second-order change. Second-order change occurs suddenly and dramatically, propelling a group to new levels of organization.

The torus attractor (Figs. 2.7, 2.8, and 2.9) used to clarify the concept of self-similarity is also a model for first-order change. Behavior in natural or social systems (marriages, groups, religious rituals, etc.) remains similar, but not identical, over time. The slight variations that do occur in any given cycle represent first-order change. These changes do not substantially change the systems; often they are hardly noticed. These small and local variations can be amplified to propel the system toward a more complex level of development, but mostly they are contained or damped and remain as minor perturbations.

Maintaining the integrity or current boundaries of any social system requires "some social response with which to defeat the transformation of the torus into a butterfly attractor. In human affairs, this requires forgiving, forgetting, and treating incompatible events as if they never happened and, in general, a continuous and expert editing of the reality process as it unfolds." So "infidelity in marriage, dishonesty of employees, disobedience of soldiers, and heresy in the priesthood"[11] are usually defined or excused as just noise in the system. My favorite example of redefining reality as just noise is teaching evaluations used by students to evaluate faculty performance. These evaluations are routinely dismissed or discounted because "the evaluation instruments are flawed; it's just a popularity contest; if you grade hard you get poor evaluations," etc. The problem gets reframed as the evaluation process, not the teacher. Therefore, constructive change is usually thwarted. In groups, one method by which positive feedback (feedback that propels the system toward reorganization) is damped is scapegoating, fully described in chapter 9.

A small change at a critical juncture can lead to second-order change, resulting in the torus breaking in two or more tori. The result might look like a butterfly attractor in which "two quite different fates" for a person, a marriage, or a group are offered.[12] "Similar systems will have different fates depending upon their journeys through uncertainty."[13] The result of second-order change in not predictable. Figure 1.10 shows a tongue forming on the torus, indicative of potential second-order change. It is important to remember that any small change can be responded to as either just noise or as a catalyst for second-order or discontinuous change. It depends on the stability of the system. In stable systems, the change is damped; in unstable systems the change becomes magnified.

Second-order change is essential for healthy group development. In fact, without these discontinuous leaps, the group-as-a-whole would never progress beyond Stage 1. However, such second-order change requires chaos. During these chaotic periods, all emotional and intellectual anchors are uprooted, plunging the group through whitewater, fraught with wild turbulence and powerful undercurrents. The group experiences what Gemmill and Wynkoop accurately described as the rampant disintegration of its current social organization. There is high anxiety, emotional distress, frustration, and anger. This disorganization is part of the process, described by Koestler, that artists undergo as they search for novel meaning.

THE VORTEX OF TRANSFORMATION

Figure 4.16 illustrates Gemmill and Wynkoop's vortex of transformation. Chaotic activity is organized around the spiral in four issue spaces representing group-as-a-whole phases that must be negotiated for continued group development. At each phase, the group undergoes a second-order change, continuing on the downward spiral, or spins off into a regressive solution and, at best, a first-order change. Change, in Gemmill and Wynkoop's model, is normative because second-order is considered more desirable than first-order.

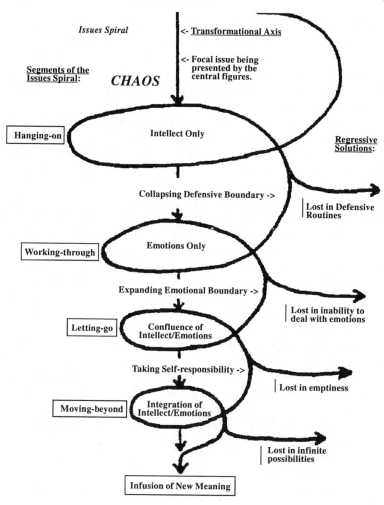

FIG. 4.16. Vortex of transformation. From G. Gemmill and C. Wynkoop (1991). Copyright © 1991 by Sage. Reprinted by permission.

Initially, the group meanders, albeit unconsciously, while central figures emerge around a new focal issue. Focal issues are created when specific emotions and attributes are split off by members and unconsciously projected onto these central figures.[14] As current group structure is found inadequate to resolve the focal issue, the group moves further from habitual patterns in seeking a solution. As critical instability is reached (a critical mass of members becomes dissatisfied), the group enters the vortex. Successful completion of the cycle results when a critical number of group members have gained insight into the focal issue. Gemmill and Wynkoop define *critical mass* as a "point at which the synergy of the 'group-as-whole' is sufficient to support the group in its movement" to the next stage of development.[15] This definition coincides with the notion of phase-locking.

The thrust of the group's movement through this vortex is collective self-awareness. Increasing awareness through each of the four issue phases makes known unconscious and preconscious issues that influence the group's behavior. As members gain consciousness, they undergo what is called the "reparation process" and reclaim their projections. With this emerging knowledge, the group learns to sustain itself toward a transformative solution. When a critical mass of members can integrate new meaning into their group, it passes through the spiral's vortex toward acquisition of the next focal issue. Then the four-phase cycle begins anew.

Gemmill and Wynkoop presented an example of a group member who expresses her boredom with the group and then storms angrily out of the room. On returning, she talks about her anger, and other members share their reactions to her. Previously hidden and unexpressed anger is now being addressed in the group. However, the anger is viewed as belonging to the individual. Were exploration of the anger to remain at this level, the group would experience, at best, a gradual or first-order change.

When members gain insight by becoming aware of what the meaning of this individual action has for the group-as-a-whole, the potential for second-order change increases. In this case, the group comes to realize that the individual member is really expressing the group-as-a-whole's boredom and anger, and her leaving reflects their own desire to walk out. Concomitant with that awareness comes more understanding about their reluctance to directly experience and express their angry feelings. New understanding of the anger's meaning transforms the group into redefining their relationships with one another.

In this example, Gemmill and Wynkoop were describing how a system is perturbed and moves from a steady state toward disequilibrium. When the member expresses her boredom, an initial perturbation occurs. It increases or is amplified when she leaves the room. On returning, it is further amplified by the group's discussion of the underlying angry feelings. Initially, the group responds to maintain the steady state by damping the perturbation, limiting its amplification by viewing the disruption as an anomaly of the individual member. Were it to stay focused on the member, it might lead to a regressive solution such as scapegoating. However, as other members become cognizant of the larger meaning of anger for the group-as-a-whole, further amplification occurs. When sufficient members evidence this group-as-a-whole awareness, a critical mass develops and propels the group irreversibly toward a new level of organization.

Once the group enters the vortex, current group structure is uprooted. In the ensuing turbulence, group members are buffeted between the transformational push for change and the defensive pull toward stability and order found in regressive solutions. Earlier, Gottman captured this dialectic process in his description of anabolic and catabolic family interactions.

Gemmill and Wynkoop likened this push/pull process to developing a muscle. "Just as a muscle must be exercised to build strength, endurance, and flexibility, members must exercise their abilities to build strength, endurance, and flexibility in handling emotions surrounding chaotic situations."[16]

When the group is inhibited from bringing preconscious and unconscious material into awareness, the result is a regressive solution. Ironically, if the group attempts to hide from its projected material, it becomes haunted by it. Just as a muscle atrophies from lack of exercise, so, too, will the group's ability to contend with future transformational possibilities atrophy if it settles for a regressive solution. If this temporary regressive solution stays unresolved, it can have detrimental effects on group members, and the group-as-a-whole can become a destructive force.

Although this example is presented in a linear fashion, in actuality the group's movement would be uneven, hesitant, and resistant to the new awareness. Undergoing second-order change requires group members to leave a comfortable, albeit unconscious, state, uproot, experience considerable discomfort and anxiety, and then reorganize. Recall our discussion of discontinuous group movement in the previous chapter. Overall group development is progressive, and creates time and history.

TIME

Throughout these last three chapters, the concept of time was introduced and discussed in various forms. The here-and-now, critical timing, flow, the group's history, and the stillpoint are all manifestations of time or our experience of time. Before concluding these chapters, I want to fit these concepts into a model that establishes a relationship among them. Let us turn then to Shlain's provocative and most original book, *Art and Physics*, for his discussion of Minkowski's space–time continuum.

The book reflects Shlain's effort to chronicle the prescient relationship between art and physics. In it, he asserts that artists, through their works, foreshadow breakthroughs in science. Read this book! Given Shlain's premise, time and its representations are a central organizing principle in the book. Before we discuss Minkowski, let us consider some of Shlain's other observations.

In his description of Marcel Duchamp's painting, "Nude Descending a Staircase," Shlain penetrates to the core of chaos. "Behind the dizzying chaotic motion on Duchamp's canvas . . . lay the idea of stillness, contained within the simultaneity of time at *c.* . . . Duchamp's Nude can be observed as existing in the past, present, and future."[17]

The letter c represents time at the speed of light. Shlain asserts that Duchamp's painting depicts Einstein's image of time at almost the speed of light. Of course, at the speed of light, past, present, and future would exist as a single image. What is represented in this picture is the inherent creative potential found in chaos; in the whirlwind produced by the merging of time's three dimensions, infinite possibilities can emerge. Everything is present; nothing is limiting, restricting, or holding back the ability of the group to maximize its potential as it reorganizes and emerges at a new level of development. Although groups entering chaos can exit in many directions, those groups who avoid chaos severely limit their growth potential.

The stillpoint is timeless, existing in a fourth dimension. Time and space issue from it. It represents the pregnant moment when all time is conjoined into a single point from which the future unfolds. It is the transition point between Bohm's implicate and explicate orders. Koestler depicts it as the "L" in his illustrations (discussed earlier) at the interface between two or more material frames of reference, the simultaneous apperception of multiple points of view. This still moment occurs just prior to the "ah-ha" experience, or leap to creative insight, experienced by many artists.

Our sense of present time is expanded when our full attention or awareness exists at the stillpoint. Past and future recede in the background. Bill Russell sensed the flow or elongation of present time during his peak experiences. Zen practice is centered on bringing conscious attention to a stillpoint.

This relativity of time is expressed in literature and film. Ambrose Bierce's entire story "Occurrence at Owl Creek Bridge" unfolds in the time span it takes a man dropped from the gallows to die.[18] The spate of *Back to the Future* movies reorients our view of time, as does the recent movie *Contact.* Minkowski's space–time light cone provides another view of the stillpoint.

MINKOWSKI'S SPACE–TIME CONTINUUM

Figure 4.17 depicts Minkowski's space–time diagram. By now you certainly recognize the spiral-like shape use to depict light cones. Minkowski was one of Einstein's former teachers and formulated the original equations that established the reciprocal relationship between space and time. He fused this relationship together into a fourth dimension that he called the *space–time continuum.* The point where the cones touch Minkowski labeled the "here-and-now." This is the stillpoint. In his diagram, the here-and-now depicts both the location of a point in three-dimensional space and the exact time it is located in that space. In this case, the exact location is the "here," at the precise moment of the present, "now."

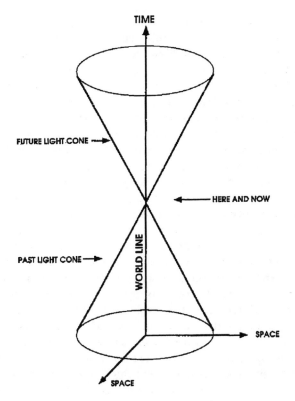

FIG. 4.17. Minkowski's space–time continuum. From L. Shlain (1991).
Copyright © 1991 by Quill. Reprinted by permission.

In the diagram, the lower cone represents the past, the upper cone the future. Minkowski believed that every object that moved through space and time left a unique history, or world-line. The light cone contains our world and everything in it. Beyond its walls exists ??elsewhere??. As Schlain acknowledges, this is the world of the contemporary cosmologist who, with the aid of the new Hubble telescope, is uncovering and mapping entire new regions of the universe. In our use of this diagram, we need not venture beyond the walls of the light cone.

From the viewpoint of the here-and-now, the observer can look into the lower cone and see the past. The farther one looks down the cone, the more past events can be observed. As one looks into the upper light cone, future possibilities exist. The closer one gets to the here-and-now, depicted by the narrowing tunnel, the fewer possibilities or options exist.

Now turn Minkowski's light cone on its side. From the point of the here-and-now, bend each light cone upward at 45° angles. Look familiar? Go back and examine the illustration (Fig. 3.10) that depicts the stage and issues of concern arcs. In the arc depicting the group stages, the stillpoint is the Conflict/Confrontation stage. From this vantage point, the group can review its past world-line created in the forming stages, and project itself into the future. At the wider end of the cone, the future appears unlimited, although immediate options or choices are restricted. Remember, future possibilities are, in part, determined by the group's history. As the group moves forward from the here-and-now, every choice or bifurcation point determines the direction of future development.

This omniscient view of the arc, with the here-and-now as the stillpoint, fixes the Conflict/Confrontation stage as the pivotal point in group development. Keep in mind, this is a snapshot; as such, it is a static picture. As the group moves in space–time, so does the locus of the here-and-now. However, in reference to the entire arc, the Conflict/Confrontation stage offers the maximum vantage point from which to view the past and the future. The arc is finite and limited to seven stages. Were it infinite, the stillpoint would always remain centered between the past and the future.

Overall group movement through the arc approximates the representation of time in this diagram. As groups collectively enter the space–time cone in the Unity stage, they are funneled toward the present. In the forming stages, discussed earlier, much of the group focus is on past experiences and events. The group leader, as we discuss in the next chapter, works to move the group into the here-and-now. As safety increases, and with proper facilitation, this movement into the present culminates at the apex in a dramatic encounter with the leader. Once arriving and fully experiencing the power, freedom, and potential of living in the here-and-now, the group continues to interact mostly in the present while progressing up the arc.

This discussion of time movement can also be used to understand and explain group movement through individual stages and activity in sessions. The space–time diagram is mirrored at each scale of group. For example, in each group session, the leader works to focus the group's attention in the here-and-now. During sessions in the descent, it mostly means shifting attention from the past to the present. In the ascent, it mostly means centering the group's attention in the present and limiting their forays into the future.

SUMMARY

In this chapter, our focus has been on nonlinear, second-order, discontinuous change that introduces novelty into groups. Without novelty, growth and development are impeded. Koestler's ideas and illustrations helped interpret this awkward notion of discontinuous change. Phase-locking was further clarified and shown as a possible explanation for how group movement is accomplished. Gemmill and Wynkoop's model of the vortex of transformation provided a vantage point from which to view a group undergoing the change process. Finally, Minkowski's space–time diagram was introduced as a framework for exploring time.

In the next two chapters, the role of the group leader is examined. The leader's role in containing group anxiety, perturbing groups toward growth, and guiding them through change processes, all while maintaining his or her own equilibrium, is fully explored.

Chapter 5

Group Leadership:
Working With Chaos

Leadership interventions are limited to two fundamental choices if you subscribe to the group development model and the ideas of change presented in the last three chapters. You can either contain the group or perturb it. The basic conviction underlying this assertion and others made in this chapter is that groups are self-organizing. Once the group forms and enters the arc, its ability to grow and develop is intrinsic. Groups leaders should facilitate the group's natural unfolding and not, as is so often the case, interfere with it. Control is an illusion, but leadership is not without power. However, the source of the power does not come from the force of the leader's intervention, but from his timing. Just as the Aikido master harnesses the energy of a stronger opponent with exact timing, so, too, does the leader contain or perturb by correctly timing his interventions.

The leader acts as both an anabolic and a catabolic agent. He soothes and provokes. During the descent, he contains the group's energy; in the ascent, he amplifies it. In each stage, he does both. In actuality, the group leader facilitates, supports, and at times, guides a process. His skills in building a safe environment, nurturing relationships, and fostering communications create the milieu in which group members organize themselves. In a literal sense, the leader "creates the space" in which the group works.

Effective leadership requires a resolute sense of self, knowledge of group dynamics, an ability to recognize group patterns, and a sensitivity to nuance. When these skills combine through practice and experience, the leader learns to act intuitively.

Golf is learned in a similar manner. An introduction to the mechanics of the golf swing and a rudimentary understanding of physics is helpful in the beginning. The golfer first learns the golf swing by consciously positioning various parts of his body relative to the golf club. Practice and constant attention are necessary. Slowly, parts of the swing blend together. As they do, the golfer's attention turns to playing the game. Playing golf combines swing mechanics with shot-making and course management. Some estimate that is takes 7 years of golf to learn all the possible combinations of shots. Course management skills require computation of distances, wind velocity, direction, grass texture, and so on.

On the course, all practice and skill is brought to bear. After assessing the conditions and selecting a club, the golfer hits the shot. However, successful execution is dependent on the golfer's willingness to trust his muscle memory and bypass any intermediate mental process. Interruption of the swing with mental reminders disrupts the natural rhythm. Over time, outstanding golfers develop a feel for the game, an ability to improvise, and an instinct for shotmaking, born out of years of practice and playing under all types of conditions. On occasion, golfers enter into the flow and play extraordinary golf without ever once being consciously aware of swinging the club.

Watch a professional hit a golf ball. The swing appears effortless. However, that swing is the product of countless hours on the practice range. Likewise, when you watch an effective group leader work, remind yourself of the hours he has spent perfecting his craft. Remember, too, that his ability was developed one skill at a time. Just like golf, effective group work combines knowledge with practice. Let us first examine the parts that constitute the effective leader, beginning with his role as container.

CONTAINMENT

Winnicott's metaphorical characterization of mothering as the creation of a "holding environment" is consistent with the idea of containment. Humphrey and Stern provide an excellent summary of Winnicott's basic ideas.[1]

Winnicott's holding environment represents the protective care the mother provides for the infant during its first few years of life. The mother's holding environment secures the child in four ways. First, she communicates her love through physical touch. Only in this manner can she convey to the young infant a sense of protection from self and the physical environment. Second, her touch provides emotional nurturance. She soothes the child when he is frightened or in pain. Third, her "holding" contains the child's normal grandiosity and aggressive tendencies. Finally, the holding environment connotes the mother's capacity to be reliable, predictable, and responsive to the child's changing needs. As the child grows, her responses must change accordingly. As the infant begins to individuate, he internalizes or introjects the holding environment. If the mothering was "good enough," the child emerges with a strong and coherent sense of self and an innate ability for affective self-regulation.

The mother–child relationship accurately depicts the group leader's early role as container. He contains the group's anxiety through his relationship with them. This containment gives the group more freedom (safety) to experience (risk) emotional growth.[2] When the group is overwrought, the leader soothes their anxiety by regulating it. The leader does not diminish or deny the anxiety, but apportions it in amounts the group can handle. Regulating overwhelming tension in the forming stages allows the group to gradually learn to regulate and soothe itself. The group cannot grow and develop autonomously without this facility.

In holding the group, the leader says to them, in essence, "I have faith in your ability to work under these conditions." The holding environment enables the group to not only experience disorder, but to appreciate it as the necessary precursor to change. The therapist communicates this holding by his closeness and willingness to be emotionally present in the group. He also secures the group by managing its boundaries.

BOUNDARY MANAGEMENT

Boundary management is the leader's effort to create a safe container. In the forming stages, the attention of group members is focused outward as they attend to their safety needs. One way in which the leader can foster group safety is to create a predictable environment. This is accomplished by providing structure, managing time and the environment, and addressing ethical obligations. Each of these areas is a basic building block in the construction of a safe and predictable environment.

Structure

During the descending stages, the leader contains the group's anxiety and uncertainty. Structured groups, those in which activities are planned and directed, alleviate some of the anxiety and uncertainty that members experience. However, prescribing the agenda fosters dependency, and members tend to rely on the leader's direction for longer periods of time. In addition, too much structure is deadening. It eliminates aliveness, spontaneity, and surprise, which are all necessary for growth. Conversely, unstructured groups place the burden of responsibility for the group's structure and direction on the members. In the latter case, anxiety is increased as the leader's control is diminished. Forming stages require a balance between structure and nonstructure, tension and safety. Members in groups where this initial equilibrium is maintained will more readily assume leadership responsibilities when the opportunities present themselves.

An important structural facet of the leader's responsibility is to provide a clear meaning and definition of the group. The meaning changes and evolves as the group organizes, but initially the leader should provide a clear vision. The meaning or vision acts as touchstone or beacon for the group during the uncertainty and turbulence of the forming stages.

Time Management

Time management means starting and ending the group at the agreed-on times. Now, that appears to be a simple, straightforward task, but the leader will be tested by group members on his ability to adhere to it. These time boundary challenges are indirect and mainly outside the awareness of group members. The primary reason group members challenge time limits is to test the leader's mettle and thus the group's safety and predictability. Can the leader be trusted to do what he says? Is he strong enough to contain our anxiety? Is he predictable? Can he establish and hold to firm boundaries? How does he handle challenges? All of these questions are answered by the leader's actions, not his words. Every action and reaction by the leader is carefully monitored by the group. One anomaly of group work is that members need to experience, firsthand, any directives the leader gives them. Leader norms are established by the leader's actions, not his directives.

For example, if the leader schedules group meetings from 6:00 p.m. to 7:30 p.m. on Wednesdays, members will test his determination to hold steadfast to those times. Only through direct experience can they gain knowledge of the leader's commitment, strength, predictability, integrity, and general fitness to lead their group. Begin groups on time. Members quickly adapt to this norm, and late arrivals are subject to peer scrutiny. If the leader permits the group to start at varying times, boundaries remain flimsy, and member concern for safety continues to be the dominant theme. One manifestation of member resistance or defensiveness in groups is tardiness, and clear time boundaries give the leader an opportunity to identify that resistance.

When groups are permitted to run overtime, members postpone working on group issues because they soon learn that the ending time can always be extended. Furthermore, the leader cannot demonstrate predictability by being lax with the time boundaries. Ending on time encourages members to begin their work at the outset of the group session.

Additionally, the leader's task throughout a group session is to monitor the time, giving members about a 15-minute warning before the group ends. This signals the group to begin preparation for adjournment by completing issues that are being discussed and to limit the introduction of new material.

Forming groups must learn how to effectively utilize their time. In the initial sessions, members may self-disclose significant material toward the end of a meeting. The leader's task is to intercept that disclosure as soon as possible, and suggest that the issue be placed first on the agenda for the next meeting.

On the rare occasions where the group is caught up in a very emotional issue and has been unable to wind down in the closing 15 minutes, the leader might use "creative" time management. If there are 5 minutes left, stretch that into 10 minutes or so if necessary. The leader should continue to push for closure while liberally permitting the plea of "1 minute remaining" to stretch into 2 if needed. This gives the group additional room for completion of their agenda while still maintaining the time boundary.

Environmental Management

The more predictable the environment becomes, the more likely members are to shift their attention from outward concerns to internal issues. This shift is similar to what I experienced when I was a runner. The route I selected dictated how contemplative the run would be. When I ran a new route, my attention was directed outward, focused on traffic patterns and hungry-looking dogs, as I attended to my physical safety. However, if I selected a route I had run many times, I was more likely to enter a meditative state. The route, traffic patterns, and dogs were all known to me. Hence, my safety concerns were minimized, allowing my attention to shift inward.

The setting is important in constructing a safe environment, too. The room should be appropriate for the size of the group, so that if members feel a need to expand their psychological space they can do so. Comfortable chairs that move with ease and that permit members to shift their positions can help moderate anxiety.

The room should be secure from outsiders. Doors should be closed so that the group space does not extend into unknown hallways. Window shades or draperies should be appropriately drawn to protect the group from outside distractions.

If at all possible, the room should have its own temperature control unit. Physiological reaction to stress often manifests itself in broad changes in body temperatures. Helping people feel comfortable is conducive to effective group participation.

The most important aspect of environmental management is using the same room for all group sessions. This adds to the external security and eliminates the need for members to reacclimate themselves to new settings.

Ethical Management

The leader is bound by ethical standards to ensure physical safety, restrain excesses, and protect the confidential nature of the group experience. Both the American Psychological Association (APA) and the American Counseling Association (ACA) have developed ethical guidelines for counselors. More specifically, the American Group Psychotherapy Association (APGA), the Association for Specialists in Group Work (ASGW), and the National Association of Social Workers (NASW) have developed ethical criteria for group leaders. The leader should familiarize himself with the ethical expectations promulgated by these professional organizations. Some of these issues should be addressed at the beginning of the first group session. These include the following:

1. *Confidentiality.* The leader is responsible for defining and discussing with members their expectations for protecting the sanctity of the group sessions. All members should participate in the discussion so that there is no doubt in anyone's mind as to the importance of respecting and holding in confidence the revelations of others. The leader should ensure that each member is given adequate time to discuss confidentiality.

2. *Keeping it in the group.* Members should be encouraged to limit the discussion of issues generated in the group sessions to the group time. Members should be discouraged from subgrouping after the session to continue the discussion of issues raised in the group.

3. *No physical acting out.* Members have a right to physical safety. The leader is charged with conveying this principle to the group.

4. *The leader's relationship with members.* A final boundary issue is the leader's relationship with members outside the group milieu. In part, this depends on the leader's job: a group therapist in private practice, a graduate student leading a laboratory group, a psychologist in an inpatient setting all have different job expectations. Nonetheless, there are some basic guidelines that can be applied no matter what the occupation may be.

The leader should not discuss group material with members outside of the meetings. There are occasions when the leader is approached by a member who is concerned about some group subject matter. That member should be encouraged to raise the issue during the next group session, rather than discuss it without the presence of other members. During the first meeting, the leader can discuss his relationship expectations with the group members, stressing that he intends to honor the keeping-it-in-the-group agreement and confine his interactions with group members to the sessions. Some members approach the leader in order to form a special relationship, albeit unconsciously, in hopes that the leader will protect them within the group. Leader vigilance against forming special relationships with any group members while maintaining regard for individual needs is imperative.

GUIDING THE GROUP

Although the leader constructs a safe container for the group and soothes their excessive anxiety, he must be cautious in providing direction. The group's natural capacity for self-organization should be fostered, and the group's responsibility for this organization should be recognized. Perhaps more than any other skill, practicing patience in the midst of high anxiety and disorder is the most effective leadership tool for promoting group development. The leader must allow the group to meander, at times aimlessly. This does not mean watching the group struggle hopelessly, but rather providing properly timed interventions that guide and encourage the group, not direct it.

Winnicott's notion of the child's "spontaneous gesture," which signals the mother that the child is ready for separation, is at the heart of properly timed interventions. As the group develops and asserts itself, the leader must recognize and affirm the members' growing independence. This reaches a climactic point in the Conflict/Confrontation stage as power is transferred from the leader to the group. This transfer cannot take place until the group signals its readiness. Introjection of the container (the leader) occurs at this stage. The group is then freed to move voluntarily without the leader.

Sensitivity to nuance, which we take up shortly, enables the leader to detect the group's signal. The transcript in the next chapter provides an example of this signaling. Before we leave this section, let us look at one other benefit of the containment.

CONTAINMENT AS CREATIVE

Building a safe container not only holds harmless the group's anxiety, it also harnesses its creative forces. Within the container's framework, the group's energies can be constrained and directed toward its development, just as the poet's creative energies are constrained by the meter or rhyme scheme of the form in which he is working. The haiku, sonnet, and limerick all have peculiar rhyme patterns that contain the poet's talents. Within these boundaries, the artist's creativity is constrained, harnessed, and, with luck, focused. In many creative activities such as writing, painting, and composing, the artist's creative energy is constrained by the medium in which he works. Containment of a group's creative forces facilitates self-organization.

PATTERN RECOGNITION

Pattern recognition is the ability to recognize and translate individual interactions to group-as-a-whole phenomena. This ability is incrementally acquired. First, an understanding of small-group behavior is necessary. Reading and studying different theorists, watching group videotapes, and testing the accuracy of one's observations with others is good practice for acquiring this skill. The second step requires the learner to be a participant in group experiences. Not only can he observe interactions firsthand, but he can also viscerally experience them, too Being a group member is the best learning milieu for the future group leader.

The final step in learning pattern recognition is leading groups. This is best done under close supervision and with a cofacilitator. As the potential leader masters acquired skills, he gradually learns to recognize the meaning of member interactions for the group-as-a-whole. He listens for metaphor, recognizes transition opportunities, and anticipates events. After much practice, as a member and leader of groups, this ability becomes reflexive. In many ways, this acquired skill is similar to that of the chess master who, over years of practice and play, learns to recognize patterns on the chess board and not just individual chess moves.

Continuing with our sports metaphors, let us consider the nonlinear dynamics of the hockey puck and the hockey player's response to it. The behavior of the puck is complex and determined by the interactions of multiple factors: mass of the puck, momentum and velocity when struck, force imparted by the stick, distance to the goal, condition of the ice, etc.[3] In spite of the enormity of computing these factors simultaneously, hockey players manage to score goals. They do it by recognizing patterns that are learned over time. Recognition of these patterns becomes so automatic that the player responds to the moving puck without thinking. Pattern recognition in groups is comparable to the identification of client complexes in individual work. However, it requires much greater acumen. Part of the reason is the level of *noise* present in the group milieu. Noise refers to the verbal and nonverbal activity of the group that occurs at both manifest and latent levels. Sorting through it for meaningful themes requires great skill. One means of sorting the proverbial wheat from the chaff is cultivating a sensitivity to nuance.

NUANCE

There is a delicate realm of perception referred to here as *nuance.* Talented group leaders exhibit this skill, and it is difficult to explain. Bohm defines it as sensitivity to the subtle meanings that emerge from dialogue with others. For him, the ability to recognize subtleties is "at the root of real intelligence."[4] Senge provides an overview of the distinction Bohm makes between *discussion* and *dialogue* that provides an understanding nuance. Discussions are interactions in which individuals are trying to influence and score points with their arguments, in order to have the group accept their opinions or observations as "correct." The intent is to influence, not to seek a higher or common truth. Meaning remains individually held; discussion, while important, is not emergent.

Senge interprets Bohm's meaning of dialogue as "meaning passing through the word," a flow between people in which meaning *emerges* as a result of collective dialogue.

> . . . a process is emergent between x and y if it is a pattern in the "union" of x and y but not in either x or y individually. More generally, a process is emergent between x and y if the degree to which it is a pattern in the union of x and y exceeds the sum of the degree to which it is a pattern in x and the degree to which it is a pattern in y.[5]

Bohm asserts that through dialogue the whole organizes the parts. Through dialogue with one another, group members are able to access a collective wisdom or "common pool of meaning" not available to individuals. For Bohm, dialogue is emergent and generative, having a life of its own that can carry groups in directions previously unimagined. Dialogue makes collective thought coherent, like the light in a laser beam. Group leaders attuned to the subtleties of dialogue can recognize and help a group access and recognize a common meaning.

Throughout the arc, the group leader's role is to find emergent meaning. Remember, group-as-a-whole phenomena are what distinguishes group work from interactions between two people. Drawing attention to the group-as-a-whole level is an early role of the facilitator.

In the descending stages, discussion occurs at the manifest level, whereas dialogue occurs at the latent level. Once issues of concern are resolved, it is possible to create dialogue at the manifest level. In the forming stages, it is the leader's role to monitor nuance or the subtleties of emergent meaning at the latent level. As the leader draws attention to group-as-a-whole issues, he is also teaching group members about the learning potential or insight that can be gained from the dialogue at the latent level. In the ascent, when mutuality has developed, members learn to recognize the value in dialogue, and it occurs often at the manifest level.

Recognition of critical or high leverage points in groups undergoing transition is sensed, rather than consciously observed. This is a most difficult skill for group leaders to achieve because most are oriented toward reductionistic rather than holistic thinking. Handling the multifaceted intricacies of groups requires embracing their complexity. To do this requires the leader to move beyond the linear, rational thinking of cause and effect and into the nonlinear realm. Ultimately, recognition of nuance occurs as intuition or a felt sense of the whole, a willingness to trust the "inner voice," the images, and the fantasies that arise from the unconscious. Unlike more reasoned interventions, responding to the group with a felt sense is more likely to entrain the leader with the group. It means learning to live with ambiguity and tolerating the attendant frustration of not knowing. In the state of not knowing, it is easier to recognize high leverage points.

HIGH LEVERAGE POINTS

A high leverage point exists as complexity, or the boundary between chaos and order. In earlier discussion, it was referred to as a bifurcation point; the stillpoint or juncture at which the shift from order into disorder, or disorder into order, occurs. Components of dynamic systems, at this point, are not quite phase-locked, nor are they chaotic. For example, ice at 32° exhibits these in-between stage characteristics: simultaneously melting around the edges and refreezing in the middle. The lightly shaded area in Fig. 5.1 depicts this edge of chaos or complexity.

In psychological struggles, it is experienced as a liminal point between maintaining control and letting go. This point is full of ambiguity, uncertainty, and opportunity. A small perturbation, at this point, can amplify the system into one state or another. Thus, if a group leader or member understands small group behavior, and nudges the group properly, at the right time and right place, that push could perturb and influence the group's movement to a new level of organization. Groups do not have to enter chaos fully to self-organize to higher levels of functioning. In fact, groups or other systems that can operate effectively at the edge of chaos

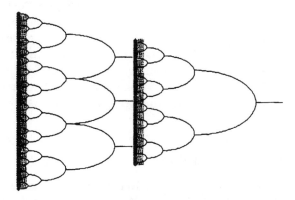

FIG. 5.1. Stability and bifurcation with complexity. From M. Butz, L. Chamberlain, and W. McCown (1997). Copyright © 1997 by Wiley. Reprinted by permission.

are highly adaptable. Adaptable and healthy functioning relationships might exist at the edge of chaos, which manifest could be depicted as the boundary between separateness and togetherness. Thus, "the repeated patterns of holding and letting go, of pulling and pushing apart, are the rhythm and breath of intimacy" in the relationship.[6]

Gifted leadership is required to maintain a group at the edge of chaos, because at the edge or stillpoint the group's self-organizing capacity is maximal. Recent research confirms that properly timed and very subtle interventions at the stillpoint can prevent a system (i.e., group) from fully plummeting into chaos.[7]

High leverage points are nonobvious.[8] They are often small, imperceptible moments in the group. They are located in group patterns and not in details, in the group fabric and not in the individual threads. Sensitivity to nuance increases the leader's probability of recognizing these opportunities.

PERTURBATION

Although the leader will perturb the group throughout its development, perturbation primarily occurs during the ascent. Once group members have moved beyond the Conflict/Confrontation stage, they are more likely to be satisfied with the group. Having survived and worked through the turbulent forming stages, the group is often content to maintain the status quo. Now the leader's emphasis shifts from containing tension to building it. Without it, the group remains stagnant. The leader can perturb, but cannot impose change on the group. An attempt to do so will only result in destructive group behavior. The effective leader can only invite, guide, push, and cajole, to perturb and energize the group.

Leader perturbation occurs in many forms. Among them are a willingness to tell the truth, to give honest and accurate feedback, and to be emotionally present in the group. The leader models behavior and asserts responsibility for what he does and does not do. He encourages group members to do the same. Depending on the group circumstances, the same leader intervention either contains or perturbs the group. Process commentary is one example. Before discussing process commentary, however, it is important to make a distinction between the manifest and latent levels of group communication.

MANIFEST AND LATENT LEVELS

There are two levels of communication that operate simultaneously in groups: manifest and latent. The manifest level refers to the group's concrete level of operation, which contains the communications and stories that are shared by the group members.

The latent level refers to the symbolic level that is outside the conscious awareness of group members and is often expressed through metaphor. It is the work area of the group-as-a-whole. Attention to this level is what distinguishes group work from individual counseling. Leaders with ability to recognize the subtle shadings of group interactions or nuance are operating at this level of communication. This is the level of metacommunication. It provides information about how interactions at the manifest level are to be understood, and defines the relationship of members and member interactions to the group-as-a-whole.

For example, two members are engaged in conflict. The manifest level consists of the overt communications between the two members. The latent level contains information about the relevance and meaning of this conflict for the group-as-a-whole. It answers questions such as, "Why is this conflict between these two members taking place now? What is the importance of this conflict for the whole group? Were these individuals elected to act out conflict for the group?"

During each session, groups fluctuate between these two levels, depending on the anxiety present in the group. Movement between the manifest and latent levels is more closely examined in chapter 8, *Group Metaphors*. Comprehension of both of these levels enhances the leader's skill, particularly his ability to observe the latent level where the therapeutic potential for change is most powerful.

PROCESS COMMENTARY OR AMPLIFICATION

Process commentary is one form of amplification. It requires the group leader to attend to both the manifest and latent levels of communication. Process commentary recycles group communication, or output, back into the system. It allows the group to continually re-examine its output or communications. Process comments are part of a feedback loop.[9] Individuals within the group work on particular tasks and issues that come about as the result of the interactions among members. This constitutes the forward movement of the group. Every activity within the group, or interaction between or among members, has potential meaning for the group-as-a-whole. Process commentary draws members' attention to the movement of the group-as-a-whole.

Commentary about group process can be fed back into the group directly, or through observations and interpretations. During the descent, when anxiety is high and containment necessary, the leader simply reflects member statements with little additional input. This is a simple iteration, feeding communications back into the group. It can amplify and draw attention to particular information. In the forming stages, group members find observations or interpretations about the group-as-a-whole threatening. Remember, group members are struggling to work out relationship issues at the manifest level. Even initial attempts to focus the group's attention on the latent level, once a safe environment has been created, may be met with silence, confusion, and misunderstanding.

Process commentary, in the ascent, can be used to perturb the group's equilibrium in two ways: observation and interpretation. As the group matures, the leader can begin offering observations as process commentary. While observations can perturb, they also teach. Mutual exploration, by the leader and the group members, of the possible meanings of individual interactions occur when attention is drawn to the group-as-a-whole, or latent, level. The leader's task in illuminating this sometimes mysterious level of group activity is to educate the members to the additional learning potential it holds for them. Once group members accept the existence of the latent level, they assume some of the responsibility for observing and commenting on it. A good method for teaching the group about the latent level is to enlist their curiosity. If the group is quiet, the leader might wonder out loud what the silence means for the group-as-a-whole. If two members are monopolizing the conversation, other group members could be asked what this behavior might mean for the whole group. If an individual expresses boredom with the group's activities, members could be asked why the group might need to feel bored at this time.

A third method of process commentary is interpretation. Interpretation of group-as-a-whole or latent level meaning should be restricted, particularly during the early sessions of group. Anxiety is increased and discussion inhibited when members fear their comments and interactions are being interpreted for underlying meaning. The leader is free to translate events during the ascent, but should not get attached to his interpretations. Sometimes the leader finds it difficult to let go of making a brilliant connection between the manifest and latent levels of the group. But there are many explanations for any one event, and some may be outside of his awareness. Leaders should not get attached to being right, but should allow for the possibility that there are multiple interpretations for any group behavior.

Continually feeding back to the group all of its output enables it to become self-reflective and self-corrective, thus supporting the group's inherent capacity to self-organize. Many group and organizational leaders attempt to control, censor, or manipulate the flow of information, which can only result in regressive group behavior. The leader is an archivalist who chronicles the group's evolution and feedbacks their history. The more skillful the leader, the more correctly timed his feedback or intervention. However, even a neophyte leader continually recycling group output can effectively energize the group.

TASK AND MAINTENANCE FUNCTIONS

Two basic responsibilities that help facilitate the forward movement of groups are attending to the *task* and *maintenance functions*. Tasking directs and focuses the group toward its stated objectives or goals. It perturbs the group. Keeping the group on task is an important facilitating function. However, it should be done in tandem with attending to the affective level of the group or maintenance function. Maintenance functions are primarily containing. Too often, group leaders become so absorbed in analyzing or deciphering a group task that they lose sight of the fact that there are human beings, with real feelings, involved in the process. Attending to how group members feel about the task at hand or about each other is critical. Maintenance lubricates the group's engine to reduce friction from group tasks and member differences. Moreover, attending to the affective level keeps members grounded in the present and helps eliminate much of the intellectualizing and abstract thinking that can distance people from one another. Disregarding the affective level of groups is one reason that many members of task-oriented organizations feel so overwhelmed, disconnected, and even discounted during group meetings.

VALIDATION AND AFFIRMATION

Two of the most effective maintenance skills are *validation* and *affirmation*. They are closely related and serve to acknowledge the humanity of each person. They can either perturb or contain, depending on group circumstances.

Validation acknowledges the value inherent in all persons regardless of their actions or behavior. For example, if a group member is having difficulty expressing a feeling, rather than push for his feelings, it is better to acknowledge the effort. "I appreciate your willingness to struggle with this; it is sometimes very difficult to express feelings." Validation is based on the belief that all persons are essentially good and always strive to do the best they can.

Affirmation is a process of making positive declarations about a group member's actions or behavior. For example:

Sam: (*to Sara*) I appreciate how you gave feedback to Jake. I've noticed several times, now in particular, that you give good, clear feedback. What you say and what you do match, and I marvel at that.

I'm really appreciative. It's nice for me to watch that. Its feels real safe when you do that with people.

The leader will discover that much more progress is made if he utilizes these two methods. However, affirmation and validation must be genuine. The group leader should not fabricate them, but should look for genuine opportunities to acknowledge the struggle, actions, and behaviors of members. Even on occasions when members appear to be sabotaging the group, searching for ways to acknowledge them may sometimes be more helpful than confrontation.

SELF-IMAGE

An effective leader must bring to the group a strong and coherent self-image. This may be an idealistic expectation, but it is one to which group leaders must aspire. Supervision and individual therapy can be a beneficial part of group leader training. Without a mature sense of self, the leader projects his own needs onto the group. The resulting countertransference distorts the group's own emerging identity and results in their inability to tolerate frustration and hold the creative tension so necessary for independent growth. They remain dependent on and emotionally enmeshed with the leader, never progressing beyond the Disunity stage. Another reason a strong sense of self is so vital has to do with transference. Transference is a process where members unconsciously project on the group leader previous experiences with authority figures, particularly parents. A common group expectation is that the leader will tell members what to do, and how to do it. Consequently, if you have eight group members, there will be eight sets of expectations about how the leader should act. With so much anticipation, the leader quickly falls short, engendering frustration and anger from the group. The various member reactions to the leader are very similar to the ways in which group members behave with authority figures in their personal lives. Focusing discussion on these similarities later in the group's development can present singular learning opportunities.

Encountering transference in the midst of forming groups can be crazy-making. Without a strong center, there will be a tendency to mishandle those projections. When a leader feels emotionally overwhelmed by a group, it could indicate that the leader's personal boundaries have been breached. Leaders must have a healthy respect for their group role and maintain a constant vigil for projections.

COFACILITATION

Ideally, cofacilitation should be the norm in leading groups. However, outside of training groups, few groups have the luxury of shared leadership. Economic concerns often dictate that schools, community agencies, HMOs, or private practices are unable to afford coleaders. Nevertheless, the process of learning group dynamics is accelerated when two leaders are paired.

There is no secret to learning to work with another group leader; mainly it is accomplished through trial and error. Classroom preparation is important, yet actual group experience is the best training format. Leaders with complementary skills should be teamed together, drawing on the strength of each. Joining a leader skilled in task-function with a strong maintenance leader provides a good working balance for the group. In addition, leaders are better able to learn in areas where they lack sufficient skill when paired with a partner who is strong in that area.

Learning when to engage and disengage with the group in accord with the movement of your partner is one of the first skills the leader learns. Cofacilitation is a process of knowing when to lead and when to follow, much like learning to dance.

Their relationship is complementary. When one leader is actively engaged with a group member, the cofacilitator should attend to the group-as-a-whole. Remember the two levels of group communication, latent and manifest. The nonengaged leader is responsible for monitoring the level opposite the one in which the engaged leader is operating. Also, keep in mind that a leader's responsibility is to involve all members in the group process.

Leaders should monitor interactions between their partners and other group members. There is always a power imbalance between leaders and group members, so it is important that all leader–member interactions be followed carefully. The cofacilitator's monitoring role is to ensure that this power imbalance is not misapplied and that members have equal opportunity to fully express themselves.

If one leader is stuck or feeling lost, he can ask his coleader for help or clarification. If he is about to engage a member of the group and wants to make sure his partner attends to the interaction, he should ask her to do so. Not only will the leader get what he wants from his partner, but he also models open communication for the group.

Initially, the safety needs of the group dictate that coleaders present a strong, united bond. As the group matures and becomes less dependent on the leaders, they are freer to work out their differences within the group.

During the early sessions, any expression of conflict between coleaders is best reserved for supervisory sessions. But as the group becomes more cohesive, it is better able to tolerate leader differences. Members can sense when leaders are in conflict; pretending that differences do not exist negates members' perceptions of reality. In addition, modeling conflict and conflict resolution can serve as a catalyst for the exploration of intermember differences.

When possible, leader pairs should be gender balanced. This pairing not only provides balance for the group, but fosters images of a family. These images, positive or negative, can be explored when the group is secure.

SUMMARY

In summary, group leaders facilitate process. They secure and sustain the group, promote the common welfare of the members, and allow the group's self-organizing capacity to emerge. In many ways, group leadership can be likened to jazz improvisation. Both the musician and the group leader must learn to create in real time. Whereas most forms of creativity, such as this book, can be edited and reworked, spontaneous activities such as jazz improvisation occur as final acts. Group leadership is also constrained by real time. Good facilitation requires immediacy. There can be no intervening mental representation or the moment is lost. When learning to lead, there is a lot of internal dialogue between the group's action and the leader's reaction as different interventions are considered. But as the leader acquires skill, his responses occur without the intermediate mental processing step, especially at bifurcation points where the timing of the intervention can be critical.

At other times, if the leader misses an intervention opportunity, all is not lost; groups often recycle issues until satisfactory resolution is reached. Thus, leaders will have other, slightly different, opportunities to intervene. The point of similarity between the musician and the group leader is in the construction of the improvisation, or intervention. Both rely on two different mental components: a long-term memory for a basic set of structures, and a set of principles that apply to improvisation.[10] For the jazz musician, that means he must know chord sequences and how to employ them.

Good leadership requires the ability to recognize group patterns, high leverage points, metaphors, group resistance, and group stages, and to know when to contain or perturb, all while remaining emotionally involved in the group. It isn't easy! The leader must simultaneously process and sort through his own emotions while observing the movement of the group-as-a-whole and its members. Just as the jazz artist learns his craft by improvising, so, too, must the leader learn to lead by leading.[11] Through practice (leading groups), he must learn to center himself, be present, and facilitate others.

WOMEN IN AUTHORITY

There is another aspect of group leadership not described in most group textbooks, having to do with women in authority. It is a particularly important issue because women are a majority in the helping professions today. As a result, group courses frequently have women cofacilitators leading all-women groups.

My training did not alert me to the differences that distinguish these groups from their mixed-gender counterparts, nor did it sensitize me to the unique issues of the female in a leadership position. No wonder female students were confused when their group experiences differed from the model I pressed them to adopt. However, after supervising all-women groups, I began to notice characteristics that appeared to distinguish them from other types of groups.

Little attention is paid to these differences in popular group textbooks. The absence of this discussion implies that all groups, regardless of composition—gender, sexual orientation, minority membership, or socioeconomic status—operate and develop in similar ways. The irony is striking. Books written from a male perspective that reinforce the societal view of power are used to train women whose experiences in leadership positions differ markedly from those of their male counterparts.

This led me to the relevant literature. I soon discovered that some others had reported characteristics that distinguished female-led groups from those led by their male counterparts.[12]

Gender-Role Formation

Gender-role stereotypes influence the expectations that group members have of female leaders.[13] Women are viewed as more accommodating, trusting, fair, and empathetic. Some suggest that gender-role perceptions are heavily influenced by family structure.[14]

Fathers are the family's representatives to the external world. Their authority derives from their connection to this outside world, and they still are perceived to have responsibility for protecting the family and providing economic support. The mother's role is relegated to a secondary position of authority, responsible for the care and socialization of the children. The mother's power is derived from her coalition with the male. When the male is absent, as in most single-parent families, the woman is seen to gain the primary authority position by default, rather than by virtue of her own abilities. In groups co-led by women, there is often a search by members for a male leader. A similar phenomenon in a work group setting with a female authority figure has been observed.[15] (A case example is found on pages 155 to 157.)

As a male supervisor of women-led groups, I notice that members, during classroom processing sessions, tend to identify me as the real group leader and to seek my advice and counsel. It is difficult to overcome this predicament, but discussion of this issue and assignment of relevant reading materials helps greatly in mitigating this circumstance.

Chodorow describes characteristics of the idealized mother that affect both men's and women's perceptions of the female: selflessness, total acceptance, self-abnegation, lack of aggressiveness and criticism, and nurturing.[16] These expectations of how women behave influence the reactions of group members toward a female leader, particularly when her actions run counter to those expectations. The terms *role incongruence* and *status incongruence* are used to describe the dilemma women in leadership positions face.[17] Role incongruence refers to the expectations that are associated with feminine roles and the resulting dissonance that occurs when women exhibit behaviors in conflict with those expectations. The same leadership behaviors that are viewed positively for men have negative connotations for women. Thus, member expectations create a dilemma for the woman. She is expected to be leaderlike, yet maintain her feminine temperament. Furthermore, women leaders are more likely to derive satisfaction from assisting others rather than asserting influence over them.[18]

Status incongruence occurs because the woman simultaneously occupies two roles: "female, which is the lower status of gender, and leader, which is a higher status position than member."[19] Members attempt to resolve this incongruity by distorting one position or the other. Since gender is a more diffuse characteristic, members first tend to discount the women's leadership ability in order to maintain their view that women have lower status. They may "act as if someone else is the leader, try to take over the leadership themselves, or behave as if the group has no leader."[20] If ignoring the leader fails, groups then ignore her gender. If all attempts fail, the group may have to reevaluate many gender-related assumptions. DeChant's *Women and Group Psychotherapy: Theory and Practice* is an excellent resource for an analysis of gender roles in groups. Works by Rich, Chodarow, Ehrenreich, Pollitt, and Gillgan are also recommended.

Group Development

The early stages of group development are further complicated by member reactions to the female leader. These reactions range from ambivalence to confusion and rebellion.[21] It is suggested that all-women groups tend to work on issues of intimacy earlier than do mixed-gender groups.[22] This concurs with my observations that higher levels of trust and safety are necessary in all-women's groups before conflict can be expressed. Inexperience in dealing with issues of power and conflict may also inhibit the development of an all-women's group.[23]

Devaluing the Leader

Another phenomenon that I have observed personally and that is reported in the literature[24] is a devaluation of the group leader. Group members covertly attempt to uncover the leader's weaknesses and then console her for having them. An example illustrates this point:

> Several members of a group were . . . in concurrent therapy with colleagues of the leader. These colleagues began reporting that their patients were expressing a wish to be "helpful" to the group therapist, since she was obviously inexperienced in group leadership and probably in motherhood as well. They confided that they could not, of course, discuss their ideas about the leader's limitations in her presence for fear of hurting her feelings. Two different therapists reported that their clients had commented about them to the effect that "she seems like a good kid."[25]

The devaluation of the therapist serves several unconscious purposes: defense against ego-alien attitudes, hostility toward the leader, and quelling anxiety about the leader's perceived malevolent power. These findings were based on the direct or supervisory experience with five therapy groups oriented toward addressing "women's issues."[26]

I noticed a similar devaluing/consoling process during observation of laboratory training groups composed of female graduate students. One of the most observable events has been the covert questioning of the facilitator's qualifications for leading groups.

In another case, a group was extremely resistant to a very competent female facilitator because she continually encouraged the group to focus on its task. The group ignored her admonitions, until out of frustration she expressed her anger. The group responded with confusion at her anger, attributing it to her misinterpretation of the group events. Finally, feeling exasperated and somewhat crazy in the face of the denial, the facilitator cried. Group members immediately tried to console her by telling her she was doing a good job.

Women's Collusion in Their Own Domination

A woman's socialization may lead to her unconscious collusion in maintaining a secondary authority role. She may conspire to maintain the nurturing role at the sacrifice of the group task. I was surprised to find, in one university department with all male faculty, that women students had been socialized to bring refreshments to the oral defense of their theses. When confronted on this behavior, it was explained away as tradition.

When women in leadership positions make reasonable task demands on the members by relinquishing their typical gender roles, members attempt to induce guilt in them. They do so by claiming the leader is "authoritarian when she is authoritative, ungiving and withholding when she is realistic, and unreasonable for expecting them to be responsible and exhibit adult behavior."[27] Further, it has been asserted that the socialization of women inhibits them from expressing anger, particularly toward males.[28] When women leaders become the objects of bitter acrimony and respond with hurt and vulnerable feelings, they often generate more anger.

The expression of conflict is difficult for many women because they have been socialized to view it as "unfeminine." Many women experience anger as hurt and disappointment, and cry or feel pain rather than express it.[29]

Although both men and women avoid conflict, women have more difficulty because of personal and social pressures. Conflict is seen as distancing behavior, and women have been socialized to value connectiveness. When conflict occurs, women leaders, like men, may experience it as a personal assault, rather than a stage-appropriate behavior, and work to dissipate it.

Several strategies are offered for effectively facilitating conflict in all-women groups. First, educational sessions alerting women to the potential benefits of power struggles and conflict inherent in groups can help members understand the importance of these issues in building effective relationships. Second, group exercises or group games that focus attention on ways in which members influence one another can be instructive. Third, skill-building exercises in interpersonal communication and feedback techniques can be introduced into group sessions.[30] In all-female groups, members move between idealized expectations of the leader (nurturance, acceptance, empowerment) to rejection and anger when the leader fails to provide them.[31] In mixed groups, women reactions to the female leader are tempered by their own relationships with males. Therefore, women in mixed groups may work to convince the male members that they are different from the leader by criticizing and rejecting her.[32]

Reaction of Men to Women Leaders

Underlying the reaction of men to female leadership is a complex relationship between the psychological development and socialization of males that takes place from birth. Bernardez captures the essence of this discussion:

> . . . many males are divorced from aspects of themselves, particularly those that are early identifications with the maternal object, by a complex process of counter-identification in response to the culture's injunction against being female-like. This negative injunction with women places men in a continuous struggle with aspects of themselves that need to be warded off, and in many cases the splitting of those aspects and their projection onto women permit a temporary equilibrium. The male thus controls in the female aspects of

himself that he fears and devalues. The domination of women is encouraged by the culture, but its strength comes from the need of males to control and dominate the female-self in themselves. This defense is threatened when a woman in power appears to have control. Not only is the male not in control of the satisfaction of his dependency needs but he may fear the loss of control of aspects of himself that are frequently projected onto women. A woman in control is thus experienced by males in this situation as controlling them, and forcing upon them behaviors that have highly negative value for them: passivity, submissiveness, compliance, dependency.[33]

It is suggested that men are less likely to confront a female leader because it is not masculine to fight with a woman. In addition, male fears of being castrated and rendered powerless have also been suggested as reasons for the avoidance of conflict.[34] When confrontation of female leaders by male members takes place, it is often indirect, subtle, and unacknowledged. It occurs as "covert defiance, denial of subordinancy, or attempts to seduce her out of her role."[35] Competition was also found to be diminished in mixed-gender groups led by women because they emphasize cooperation.

Women leading groups must not only address issues of conflict and confrontation attendant with the group's growing independence, but must be cognizant of gender issues, too. Gender issues must be addressed when they appear in the group. Just as indirect anger is brought into the group's awareness, so, too, must circular references to gender issues. Sensitizing groups to these issues can help ameliorate them. Furthermore, they can serve as a catalyst for explorations of gender issues among members. Finally, keep in mind that stage development in all-women groups are slightly altered. Women develop more group cohesiveness before they enter the conflict period.

Chapter 6

Group Leadership:
More Skills

There are specific leadership tasks that need to be undertaken at different points in the arc. Although some are stage-specific, many continue throughout the group.

MODELING

In beginning sessions, members carefully monitor the actions and reactions of the leader to events in the group. Keep in mind that the group is just beginning to formulate its norm of behaviors. As the authority figure, the leader is watched for clues to appropriate behavior. Members also believe that the leader can predict with a degree of certainty what will happen in the group.

CLEAR COMMUNICATION

Because of this monitoring, the leader, through example, can help establish some effective methods of communication. She should use "I" statements when she speaks, taking responsibility for what she says in the group. "You" statements should be avoided, as well as "we" statements, when working with a coleader. Speak for yourself so that group members learn to speak for themselves.

AFFECT

If the leader makes an affective statement, she should be sure it is a feeling that is being expressed. Too often abstract words such as "good," "fine," and "okay" are substituted for expressions of affect. Frequently, an "I feel" stem is followed by a thought or observation. Members avoid the sometimes difficult task of identifying feelings if they observe the leader also avoiding them.

Time should be taken when expressing affect to identify exactly what is felt. Sometimes, if the leader can just identify the physical sensations present in her body, it can serve as a catalyst for discovering the underlying affect. A leader's request for a feeling response from a member is strengthened if she has worked to identify her own.

ANXIETY

Every group leader experiences anxiety when beginning a group. Her feelings about the new group environment are no different from those of the members. However, during the initial sessions, she should hold back from sharing her discomfort with the group. This does not deny her feelings of anxiety; it simply puts expression of them on hold.

I hesitate to make this statement because I firmly believe that honest sharing is crucial for effective group leadership. However, the overriding concern in the opening group sessions is for the leader to demonstrate confidence, even if confidence is not what she feels. If the leader tells the group during the first session that she is scared and unsure of herself, such a statement, although true, may increase members' already high levels of anxiety. Beginning groups need to feel safety; much of that is generated by the leader. Young children depend on parents for their initial safety needs; likewise the group depends on the leader.

My reluctance in suggesting that the leader reserve expression of her anxiety feelings for supervisory sessions stems from a concern that this will be read as an admonition against leaders sharing their feelings. That is not what is being advocated here. The group is best served in the early sessions by demonstrations of the leader's competence, not her uncertainties.

Thus, in the beginning it's important for the leader to demonstrate confidence and guidance. Providing an agenda for at least part of the group's first session can accomplish this goal. An introductory exercise or other structured activity can diminish anxiety and regulate the amount of self-disclosure from each member.

Leadership in this forming period has considerable import for the group. Much of what the leader says and does helps construct the early group norms. Perhaps at no other time in the group process, with the exception of the Conflict/Confrontation stage, will the leader be as closely monitored as in this initial period: Too much anxiety will overwhelm the group, too little will inhibit its movement. At important junctures in the initial sessions, the leader should exercise caution. Cautiousness is a virtue in facilitating forming groups. The risk the leader takes by increasing anxiety (e.g., by requesting more self-disclosure) is to frighten members into defensive postures from which some may never recover. Some early casualties, members who leave groups after only one or two sessions, may have fallen victim to such leader enthusiasm.

THE NORM OF SAYING "ENOUGH"

An important norm that should be established early in the group is one that gives members permission to control the amount of their self-disclosure. Frequently groups may single out one member, in order to lessen anxiety, who does much of the talking during the first session. Members keep attention focused on this person by asking him questions. Therefore, other members are protected from having to self-disclose.

The talkative member is placed at risk because no norm has yet been established for shifting the group's attention when someone feels overloaded. The leader's role is to help members establish that norm. In the following example, notice how the leader (Brad) helps one member establish that norm for the group.

This transcript was taken from the first session of a graduate training group. Some members had been sharing bits of biographical information when Becky stated she'd like to hear from everyone in the group. Notice how Brad (the leader) helps Marion establish the self-disclosure norm.

Brad: (*to Becky*) Is there someone in particular you'd like to meet or hear from?

Becky: (*laughs, pointing to Marion*) I'm kind of interested in Marion. I don't know Marion at all. (*To Marion*) You just seem kind of neat to me. I don't know why, but—

Marion: Well, I'm from Woodsdale. I grew up here. I have a B.A. degree in psychology, and I just started this program. I'm looking for a job in my field. Today I applied for a job down at Woodsdale Mental Health Center. I am waiting to hear from them in a couple of days.

Becky: What position would that be?

Marion: Mental Health Assistant.

Barb: Are you pretty anxious about the job application?

Marion: I'm not real anxious. I have a pretty good possibility of getting hired. My mom works down there, and I know a couple of the people who work down there so that— I don't know if that has anything to do with getting a job or not. Also I have some good qualifications, too, to be able to do the job.

Marilyn: What kind of things would the job entail?

Marion: Hmmm. It's just basically being there for people—ah—answering the phones, and just like going around and checking on people, and see how they're doing, and talk to them, and just basically just kind of a counseling position. You don't have a title, but it's informal. You answer the phones. I'm not really sure of everything I will be doing.

Hank: You got any job experience? Are you just out of school? Just graduate?

Notice how the group moves to relieve the anxiety by exclusively directing attention toward Marion. However, if you ask groups members about this experience, they would probably feel they had all been equally involved in self-disclosing. Much of what is happening here is outside of their awareness.

Marion: Just out of school.

Hank: Just graduate this spring, and now you started this.

Marion: I've had some informal training working at Woodland in town here. It's an abusive treatment, or abusive safe home for battered women and children. I have done some things there in the past.

Hank: Would you like to stay in this graduate program even if you get this job?

Marion: Hopefully, if everything works out.

Hank: What would you like to be doing? Would you like to be a school counselor now?

Marion: Hmmm, eventually, yeah. I'd like to stick with what I'm doing now, working with women and children that have been abused because I've had a couple of years' experience with it and I am just getting a good personal knowledge of it. It takes a real long time to get a really good feeling of what these women have been through in their lives. So I like to stick to that for awhile.

Hank: It is real stressful to work with these people?

Marion: Sometimes it is. Sometimes, ah, it's annoying, especially because as a volunteer I'm there, and I, I'm a crisis phone worker. I answer the phones a lot, and women will call up and just want to—want to remain anonymous—just want to talk about their problems, and it's real annoying to sit for an hour and listen to someone and it's also frustrating to try and be a helping person over the phone and just not getting a lot of things done.

Hank: Why is it important to get their names?

Marion: Well, we're there to help them, but a lot of times they're
           afraid to start anything, they're afraid to step forward and
           admit that they have been in an abusive home. And, they're
           afraid to do anything about it. We're just there to help them.

Hank: Well are you in a position if you found out their name would
           you call them back?

Marion: Hmmm. It's up to them, it's their choice to come, but we
           can't force them to come. . . . I guess it's their choice.

Hank: So the frustration lies more with that they don't want to do
           anything to change their life and you feel frustrated because
           you can't get them to take a step forward. Is that it?

Marion: (*restlessly*) Yeah, well just the environment in which a lot of
           the—just over the phone if they call in—and just being a
           volunteer and the hours that I work, too. I can't, I can't do
           a lot of stuff the regular staff people do. So I always have to
           make referrals and—well you have to call back tomorrow and
           talk to someone else, because I can't do anything for you now.

Hank: So it's kind of frustrating not being able to follow through,
           or complete. You start the ground work and you just have to
           turn it over to someone else.

Brad: (*attempting to shift the focus from Marion, and move it into the
           here-and-now*) I'm kind of wondering what kind of frustration
           is going on in the group right now, if any. I just want to check
           that out.

Group: (*Silence*)

Brad: I am really wondering what is going on. What's happening
           right now in the group?

Mark: I feel that this is a tension reducing situation for us. We focus
           on something that's sort of in the past and that's outside of
           the group and it's a little more comfortable. It's concrete and
           it's easy for us to talk about.

Brad: (*trying to get Mark to own his statement*) It's more comfortable
           for you, Mark, to have that happen?

Mark: I don't know—maybe not for me, but I'm wondering if that's
           not the whole process, as far as—I'm not sure if it's more
           comfortable for me. (*shifts in his seat, voice is shaky. He appears
           nervous, and so shifts the focus back to Marion*) I guess I'd like to
           hear more about how you feel now about the people you are
           dealing with. People who weren't able to leave their situations,
           weren't able to get more healthy.

Marion: I feel like maybe I haven't done the best job I could have—
Bill: (*interrupting*) You felt a certain responsibility?
Marion: Yeah. I feel responsible and I have these feelings that go with
         me after, after it's happened. Thinking I should have done
         this, or I could have said this. Maybe something I would have
         done would have facilitated them to come in. So I guess my
         feeling now about that situation would be that I maybe feel
         a little bit of guilt or not being adequate.
Mark: A little doubt.
Brad: (*intervenes as container*) Marion, I'm wondering how you're
         feeling right now about being the center of attention for this
         group?
Marion: I feel pretty comfortable. I feel like I'm kind of intellectualizing
         my talk though. I feel that we maybe should get to something
         here-and-now, and not talk about things that I've done, things
         that I have felt.
Brad: Does it feel like too much focus on you?
Marion: I think it would be in a few more minutes, probably.
Brad: (*Here, the leader wants to shape the norm*) How would we know
         that?
Marion: Uh, I suppose I could bring it up.
Brad: Say, "I've had enough," or "stop" or—
Marion: (*interrupting*) Yeah! So I've had enough!
Group: (*laughter*)

It is very important initially to help members actually phrase the words
for this norm as Brad has done here. Simply leaving it with Marion stating
"I suppose I could bring it up" is incomplete. As you notice, almost simultaneous with Brad suggesting a phrase, Marion responds. Once this norm
is established, safety is increased. Members now have permission to take
care of themselves.

Keep in mind that members must directly experience the establishment
of a norm. Prescribing the norm is not as effective as helping group members establish it for themselves. The leader should look for an opportunity
to form this norm in the first group session. It increases feelings of safety
and gives members the responsibility for protecting themselves.

As safety increases and norms of behaviors are established, the group
begins its indirect challenges of the leader. In the initial sessions, members
often challenge the leadership, particularly when anxiety is high. The
challenge occurs because members believe that leaders are responsible for
controlling the anxiety in the group and making them feel comfortable.

The challenges are subtle and indirect, often appearing in metaphors. The following example occurred during the first session of a new group. The group was co-led by an experienced leader (Sally) and another in training (Mark). This is a clear example of an indirect challenge, but the leader's response perturbed rather than contained the group. Since it was the first session, Mary and the group were pushed back.

Mary: . . . a couple of weeks ago we were in Arizona. I was looking at a program there called Family Studies, because my husband was thinking of taking a position at the university there. . . . While talking with them I discovered that if you want to go into Marriage and Family Counseling you don't even need a graduate degree. All you need is your name on the door with a little family counseling or marriage and family counseling written behind it. And that really shocked me. To me is seems unethical to be counseling somebody without any kind of training background. What do you all think?

Mary had made several statements similar to this one in the preceding 20 minutes. Several times her gaze was directed at the inexperienced leader, as if challenging her.

Laura: I don't think you need a degree that shows you're able to help someone or able to give advice or share feelings. I don't think you need a degree to do that, but I do think you need training in counseling and knowledge of different places where you could refer them. . . . You need knowledge of the city area, and the facilities they have.

Sally: (*focusing in the here-and-now*) I'm wondering if in some way you're questioning whether the facilitators in this group— Mark or me—are qualified to lead.

Mary: (*quickly and emphatically*) Oh, no! Not at all! I'm a—I see myself as a really direct person and if I had that kind of a problem I wouldn't go about it in a roundabout way. I would ask you directly.

Sally: So you don't have any concerns—

Mary: (*interrupting*) Oh no!

Sally: (*continuing*) About either Mark's or my qualifications?

Mary: (*nervously*) I think you're highly qualified.

Mark: I wonder that, too

Mary: (*puzzled*) Oh really.

Mark: . . . Oh yeah. I was thinking that, too. And I was thinking in particular—was wondering if you were comfortable because I—she has her doctorate and I just have my masters degree'. I was wondering if—

Mary: (*interrupting*) No, not at all. It was a thing that came up a couple of weeks ago. It was really a shock to me, to think that somebody would be in a counseling practice. I'm not even thinking of this particular one—I'm not thinking of an educational environment . . .

In this example, Mary and the group would have been better served if her comments had been left alone. It was much too early in the group, and too unsafe, to bring this issue into the group's awareness. The leader's interventions only served to frighten Mary and push her away. A more appropriate response might have been to simply offer examples of the leader's qualifications or to acknowledge that pairing an experienced and an inexperienced leader protects the group. Either of these observations could have been offered at any time. However, to keep anxiety manageable, offering them somewhat later after Mary's remarks might be prudent. When the group is safer, and higher levels of trust are established, it is more timely to bring into awareness any potential frustrations and anger with the leaders.

As the group matures, members try out the leadership role. Generally, indirect power struggles occur among members and between the leader and members in the descent. However, on occasion, a member may find herself thrust into a leadership position. The following example illustrates the difficulty members have in accepting the leader's role prematurely.

The group was in the Disunity stage. In the previous session, Molly had interceded very effectively between the facilitator and another group member. Molly intervened partly because she felt the leader was too challenging of another member. For a brief period of time, she took charge and led the group. She began the following session.

Molly: . . . I feel like things went too fast for me yesterday, and maybe for the whole group, but it felt for me too much. I like—I said it felt real burdening whatever was going on so I was just kind of feeling—like—I don't want to pull back today, or just—I don't want to pull myself away from the group. . . . I feel a part and want to see what's going on but I just don't want to get out there.

Kent: (*coleader*) It felt like you were really out there yesterday and don't want to—

Molly: (*interrupting*) Too much out, and taking too much in for me. Feeling too much responsibility or something for the group . . . and I did feel like I was starting to be the facilitator, like taking on—trying to take your job on.

Kent: You don't want it?

Molly: No, I don't. (*nervous group laughter*)

Kent: (*smiling*) I don't want it back.

Molly: (*laughing*) No, I don't want it.

Kent: Who would you like to have it?

Molly: The job of leader?

Kent: (*nods*)

Molly: I would probably like you and Nancy to have the job, because you are the leader (*pause*) and I would just like to kind of settle back today and be more of an observer in the group. You know kind of just—(*sighs*) you know not try and take on so much stuff. I don't think it's good for me.

Nancy: (*coleader*) Would it seem more orderly to you if Kent and I took on the role and there would be less anxiety and less tension for the group? Took charge. I got a sense when you were talking we're supposed to take charge. We should somehow be doing more, are you thinking that? Feeling that about us?

Molly: Yeah, somewhat. Although, I feel like that's possible that you should, and yet I can see my part in that. I mean my taking on for some reason, something inside me made me want to take it on. . . . I want you guys to do it. And I guess, I guess that's what happened yesterday. . . . I didn't feel protected myself. I thought that was your job or something (*pause*).

Nancy: Do you have any anger or frustration toward him (*Kent*) that he hasn't taken on—or toward me, that's two questions.

Molly: I remember feeling more on the way home last night or as I was making sense of it all. But then too even at the group time . . . you (*Kent*) owned up. You said "you're right I was pushing Marla."

Kent: . . . You'd like me to be responsible for the members here so that it doesn't go too far?

Molly: (*nods*) un huh.

Kent: And I'm responsible for the boundaries of each person here?

Molly: Yeah.

Kent: I wonder how the rest of the group feels about that?

Ali: My first thought when you say that is that I've seen you do that already with Marla. I've seen it several times you said "well how are we going to know when to stop or how are we going to know when you're feeling bad?" I guess in that respect I see you doing that right now. (*to Molly*) He didn't do that yesterday with you, but I guess I've seen that happen, so I guess in that sense he has protected or has taken steps to do that. But I'm not sure if that's what you mean or not?

Molly: Yeah, I mean—I can see that, too. Yea, I think he has. There were just instances where I feel I take it on myself.

Kent: (*validating her actions*) It seems to me you handled it quite well yesterday when you took that role (*leader*) on yourself.

Molly: Yeah, ah (*pause*) but I guess yesterday that's what I was feeling. I was taking on more than I really wanted to at this point. 'Cause I'm, you know, it just seems like too much. I have too much, just everything felt like a burden to me yesterday, and I went away thinking I've got enough to do without worrying about or taking care of everybody in this group and that's where that came from.

Notice in this example how Molly had replaced the leader in the previous session and assumed the leadership position. Her struggles come from the realization that leadership means responsibility. In this session, she wants to return to being a group member, allowing the facilitator to be responsible for the group. She is also struggling with dependency issues. Nancy, the coleader, invites Molly to express her anger toward Kent (*coleader*), but she and the group are not quite ready. However, when anger appears in this stage in any form, the facilitators should invite its expression. The leader's skill, acumen, and mettle are tested once the conflict period is entered.

CONFLICT/CONFRONTATION STAGE

Confrontation over leadership is essential for group growth. Bear in mind that the anger and frustration is directed at the leader's group role and not at her personally. Certainly she experiences the attack on a personal level, but remember, members are frustrated with her inability to meet their leadership expectations.

As member expectations go unmet, the level of frustration and anger in the group builds; without adequate expression, the group's development ceases. Many groups never advance beyond this point.

ATTACK ON THE LEADERSHIP:
MODELING THE EXPRESSION OF ANGER
AND CONFLICT RESOLUTION

As previously discussed, attacks on the leadership are over issues of power and control, dependence and independence, and transference. The group's frustration with the leader's inability to meet their expectations, coupled with their counterdependency needs, is a prime mover in the eventual confrontation with the leader. During this period, members must confront their capacity to lead.

Norms for the expression of anger and conflict resolution established during the Conflict/Confrontation stage are authored by the leader. As the frustration over leadership increases, expression of indirect anger occurs. The metaphors that members bring to group often reflect their discontent with the leadership.

Keep in mind that no matter how effective the leader is, she will not satisfy the expectations of all of the members. Her task, then, during this period, is to encourage members to voice their anger. This is a difficult, but essential, task. Generally, groups avoid conflict. They withhold, deny, and diminish their anger. Nevertheless, once the group has moved beyond the initial sessions and a minimum level of cohesiveness has been established, the leader must bring about the direct expression of anger. Once anger surfaces and can be tolerated and expressed, the attack on the leader begins.

Remember, the basis of the attack is the leader's role in the group, but interestingly it is also connected to some blind spot in the leader's personality. The group may feel the leader is uncaring, too emotive, too distant, not strong enough, and so on. The issue they choose has a basis in reality, so the leader must remain open to the feedback while remaining cognizant of the developmental necessity of the conflict. Initial reactions by the leader might be defensive. She might feel unjustly accused, hurt, or angry that the members haven't appreciated her efforts on their behalf. However, it is the leader who has empowered the group to relinquish their dependency on her, by constructing a safe enough environment in which to challenge her. During the conflict period, the leader should remain centered by expressing her feelings. The group can handle it. This is a critical time. When attacked, the leader must push back. As the group asserts itself and takes charge of its own destiny, the leader must remain steadfast in the face of their challenge. Her role must not be relinquished without a fight. It is only through engagement with the leader that the group can directly experience its own power.

Members must prove to themselves that they are capable of leading. The most effective means by which they can apprehend their power is by overthrowing the leader. Members must earn the mantle of leadership by dethroning the king or queen. So the leader must remain firm and engage the group in a straightforward manner. Actualization of the group's power rises from the heat of conflict while simultaneously strengthening the cohesiveness among members.

The battle ends when the leader sheds a little blood (*figuratively*). Bear in mind that until this point, group members have unrealistically characterized the leader as superhuman. Her earlier roles as container and protector were to ensure member survival during the anxiety-filled beginning sessions. Now members must see her more realistically if they are to govern themselves.

Shedding her blood means revealing her humanness, through the expression of feelings, and perhaps, some personal disclosure that demonstrates her fallibility. This moment represents the stillpoint in the arc. When successfully navigated, the group power base shifts from the leader to the members.

Coleader's Role

If there is a coleader, he must facilitate the attack by monitoring the interaction between the leader and the group member(s) involved. The importance of this cannot be overemphasized, because the norm for future conflict-based interactions is forged here.

Attention to the affective level is essential. The all-important question, "How are you feeling (about what is happening)?" must be asked. Moreover, a feeling response is needed from the individuals involved before the group can continue. This is the coleader's role to ensure that all individuals (including the leader) involved in the confrontation are connected to their feelings. As a protective mechanism during this turbulent time, it is sometimes automatic to disconnect from one's feelings. Time must be taken to allow everyone to feel the impact of the encounter. Many groups get stuck at this point!

There are three reasons for identifying and expressing feelings: one, to ensure that each member is fully aware and experiencing the ongoing interactions; two, so that everyone can observe that neither the leader nor the member(s) involved were hurt by the encounter; and three, to demonstrate that group members are strong enough to tolerate conflict and, by assumption, capable of handling any other difficult issues the group may confront.

After the confrontation, all members should be addressed to determine their affective states. This may seem a bit tedious, so let me clarify its importance. The reason for ascertaining feeling levels from each member is to ensure that they are processing their feelings about the interaction.

During times of strife, some members may disconnect from their feelings. Then at some later period, usually long after group has concluded, they experience that affect. The (co)leader's role is to help members feel them now, so that everyone can process their reactions with one another and complete this experience as close to the actual event as possible.

Feelings generated by conflict are usually frightening, so individuals tend to suppress their expression. The leader must remain patient. This stage in the group's development cannot be rushed. The conflict resolution norm that gets developed here becomes the model for all future conflict-based interactions in the group. The leader should proceed slowly! The group learns during this period that it can survive disunity, anger, even hostility. Even though it is frightening, group members come to realize that the group did not disintegrate during conflict.

A common fear associated with the expression of anger is that it drives people away. What members experience is that direct and honest expression of feelings, even anger, draws people together. Once the expression of negative feelings has been accepted, group members feel safer to share more intimate feelings. The attack on the leadership may take place over several sessions. The leader must commit to work in this stage until the conflict is resolved.

Attacking Both of You

In situations where there are two leaders, the group may attempt to lump them together. This must be avoided. When coleaders notice that group members are referring to both of them as "you," one leader must intervene and have them speak directly to the other. If both leaders get caught in the conflict, no one will be available to facilitate it. Groups want to resolve relationships with both leaders during this stage. It should be done one leader at a time.

Member Reactions

During the attack, members' roles often mirror their previous life experiences with anger. Some are confrontative and lead, others actively follow, and a few may mildly attempt to defend the leader. In many cases, the stance they take can be directly traced to their previous experiences with authority figures, most notably parents. All members, even the silent ones, are engaged in the attack to some extent and invested in the leader's eventual overthrow. Members who are more familiar (notice I didn't say comfortable) with expressing anger usually lead the confrontation. The predominant feelings during this period of chaotic activity are frustration, anger, and confusion.

Leader Feelings

During confrontation, the leader may experience a great deal of self-doubt. She may question her competence and even her desire to lead groups. She may even feel failure. Her self-esteem may be bruised. These are common feelings experienced by most novice (and many experienced) leaders at this stage in the group's development. This is difficult. Even though a leader can understand how necessary this crossroad is in the group's evolution, it is still painful.

Throughout the stormy period, the leader might be confronted with many personal issues that are unclear and unresolved. However, what is paramount are her feelings about anger. Is she comfortable with the expression of anger? Can she express anger clearly? In many cases the answer will be no. What to do then?

The first course of action is to address this concern in supervision. Some questions to be answered are: How and from whom did the leader learn anger? Was previous anger associated with negative events? How does the leader feel about engaging in conflict? What happens when conflict arises in relationships with friends? Family? Is it ignored or diminished or smoothed over? How did the leader's parents deal with anger?

These are questions that need to be discussed with a supervisor. Not only will the leader have to confront her feelings about anger, but many other feelings will arise during this tumultuous period.

She may find herself angry with certain group members, cool toward others, and drawn toward some. Often these feelings are based on countertransference. Her reactions to some members are triggered by feelings that may have occurred with similar persons in her personal life. Singer categorized countertransference behaviors into three groups: "reactions of irrational 'kindness' and 'concern'; reactions of irrational hostility toward the patient; and anxiety reactions by the therapist to his patient."[1] He noted that these reactions could occur either in a waking or dream state. Other kinds of feelings may occur throughout the contentious period.

Confusion. The leader often experiences a lot of confusing feelings, and how she handles them affects the group. Supervision is necessary, along with her willingness to openly address these issues. The more unclear and muddled her feelings are, the less able she is to lead the group.

What does the leader do in the group when she is confused? She must center herself, by commenting "I feel confused." Surprisingly, this simple statement helps get her back in touch with the group. The danger occurs when she attempts to suppress her bewilderment, which only leads to more confusion. Her energy becomes focused inward instead of on the group's activities. It is better to opt for expression rather than suppression.

Part of the difficulty the leader may have is exposing her underbelly by acknowledging that she doesn't know what is happening. Keep in mind, it is impossible to attend to all the simultaneous action that occurs in groups. Leaders don't have to be experts, or right, but they do have to be emotionally present in the group. Besides, just getting present can take a lot of work. When the leader has difficulty focusing on what the group is doing, it usually indicates that some blind spot in her personality has been activated.

Boredom. A common sign of avoidance behavior is boredom. It can signal resistance to some aspect of the group's activity. It might also indicate that countertransference feelings are present. The quickest way to let go of boredom and find out what is behind it is to express it. "I feel bored." Expression of boredom in the group is appropriate, but the place to explore countertransference is in supervision.

When the leader is emotionally present and the group is operating in the here-and-now, boredom seldom exists. The leader often helps the group get present by addressing her boredom. A process comment such as, "What are we avoiding?" helps refocus the group's attention. Group boredom usually masks an issue that is too sensitive, too unsafe, for the group to address. If so, pay attention to the metaphors for clues and directions on how to proceed.

Tiredness. One of the most frequent reasons beginning group leaders give for their inability to stay present in the group is tiredness. Reasons such as, "I had a busy day; I didn't get much sleep last night; I had to work all day," are the most commonly heard justifications. Although each of these reasons may be partially or wholly correct, the leader needs to explore the possibility that she may be avoiding something in the group.

Is there something she doesn't want to address? Is there a member she is avoiding? If the leader can remain present in the group, she finds it a very energizing experience. Oftentimes the tiredness she brought with her quickly dissipates. Tiredness should be viewed and considered as resistance. Even if it turns out that the leader was just plain worn out, the group is better served.

Fear. Another feeling associated with leading groups, particularly in the forming stages, is fear. Fear is common. The leader might be reluctant to pursue the feeling for fear of uncovering something dreadful. I have yet to see a group leader uncover anything dreadful. There certainly are some very painful issues that have surfaced, but never anything that was debilitating. Besides, it turns out that letting go of the fear is quite an exhilarating experience.

The leader's willingness to address personal issues both inside and outside of the group has a positive impact on the group's progress. As the leader expands her personal boundaries, so, too, does the group.

Stuck. Most often when the group is stuck, something is not being said. Someone is withholding something from the group. No matter how the group tries to resolve this impasse, it cannot and will not until the hidden issue is made known.

When the group is stuck, the leader should work to stay present (centered). Metaphors can provide information about the group's circumstances. The group should be encouraged to confront the impasse by asking "What isn't being said?" "What am I withholding?" Scapegoating may occur and draw the leader's attention from the real issue. The leader must examine her role and talk with her supervisor about providing answers to questions like these: What is she unwilling to share with the group? What is inhibiting her? Whatever the issue, it affects the group's level of functioning. The leader should work to resolve it.

Being Right. When group members attack the leader, she might find herself locked in struggle trying to convince members to subscribe to her view of reality. If so, it is a no-win situation. Being right consumes a great amount of energy at considerable expense to the group. There are always competing views of reality, each one quite legitimate. Being right is about proving someone else wrong. Quite frankly, who cares who is right?

When this occurs, the leader might be covering up some insecurities about what is happening in the group. It is more honest to be insecure than to be right. If she finds herself invested in being right, she should step back and disengage. Let the other person be right. She can offer process commentary on the dynamics of the power struggle with her energy. When groups remain stuck, a member might be singled out for blame. This can occur as scapegoating.

SCAPEGOATING

There are at least four situations in which the scapegoating response is evoked. First, feelings of frustration and anger at the leader may escalate within the group before appropriate norms for their expression and resolution have been established. The group usually directs those feelings toward one of its members. This occurs early in the forming stages. Second, authoritarian leadership, the style in which the leader imposes her will on the group, often elicits a scapegoating response. Third, groups in which there is unresolved conflict between the coleaders usually results in a scapegoated member. Fourth, incongruence between the leader's verbal and nonverbal behavior may also trigger scapegoating behavior from the group.

Scapegoating serves to focus attention away from the source of discrepancy or conflict. The scapegoat is often selected based on minimal information. This can be any incident that the group dramatizes out of proportion to the actual event.

The member who is selected as scapegoat often occupies this position in his family of origin and is quite familiar with the role. The inability of group members to comment on this dysfunctional behavior in the beginning sessions is similar to the inability of young children to comment on their parents' double-binding communications. Parents, like group leaders, are imbued with survival value.

When scapegoating occurs, the focus of attention is constantly directed at one member. The leader may unconsciously collude by facilitating interactions between that member and other group members with little regard for the scapegoated member's feelings. An inability to empathize with the scapegoated member should signal the leader that some blindspot in her personality may have been activated.

Scapegoated Response

The scapegoat is unable to understand the nature of the anger. He feels overwhelmed and confused by the group's attack, and his confusion elicits further attack by other group members. Without intervention, this can have damaging psychological effects on group members. Several years ago, I was asked to observe a few sessions of a training group that was stuck and was experiencing, I was told, strong resistance from one of its members. For 2 weeks I observed the group in action from behind a one-way mirror. It was readily apparent to me that one of the members, Judy, was being scapegoated.

The members were angry and frustrated with her inability to express feelings about the group experience, which they interpreted as fear. Her response to their inquiries and badgering was that she felt fine. However, her actions belied her statements. She sat with her legs pulled under her. Her voice was shaky, and she avoided eye contact with the others. Of course she was afraid; the entire group's attention was focused on her!

The two leaders, Peggy and Tim, participated in the badgering. There was no evidence of support for Judy, nor was there any apparent understanding of her position in the group. The group would attack, demand that she express a feeling, then retreat out of frustration. Judy was feeling crazy, and was so defensive that it was impossible for her to risk sharing any feelings with the group.

I met with the leaders after two sessions, and pointed out to them what I thought was taking place. They were dumbfounded. I focused our discussion on the relationship between them, and after some effort uncovered an area of conflict.

Peggy was frustrated by Tim's inability to demonstrate a firm resolve. She felt he was always backing down in the group, changing his position, so that neither she nor the group ever knew where he stood or what he was feeling. He never showed any backbone. Peggy felt she had to be strong for both of them.

Tim became angry at what he characterized as her dictatorial manner. He felt she was inflexible, and he was tired of her constantly monitoring his behavior. The discussion became heated as Peggy and Tim confronted their feelings about one another. After airing their feelings, Tim and Peggy were more understanding of each other's leadership style.

In the course of the ensuing conservation, Tim revealed that he was intimidated by Peggy's style and felt fearful of her. He gave specific examples of her behavior that colored his perception. She was taken aback that those behaviors could be construed in such a manner, but remained open to the possibility that his feedback was grounded in reality. They agreed to discuss their encounter with the group.

During the next session, they shared what they had uncovered with the members. There was a noticeable difference in the group; even Judy appeared to relax, stretching out her legs. As the underlying conflict was exposed, her scapegoated role was no longer needed. Each of the group members had an opportunity to learn from the experience.

The suppressed conflict had been deflected onto Judy, a role I suspect she occupied at other times in her life. The group focused on Judy as if it was her fear that was blocking it from continuing. The projection became a self-fulfilling prophecy as Judy became fearful of the group's accusations. Once the conflict between the leaders was exposed, the group was free to openly examine their dilemma.

If the leader believes that a member is being scapegoated, she should ask herself the following questions: "Am I avoiding anger? Am I directing the group too much? Is the group afraid of me? Am I afraid of the group's confrontation? Do I have empathy for all of the group members?" Most important, she should talk it out in supervision.

WHEN THE CONFLICT STAGE ENDS

The encounter with the leader ends after revelation of her humanness. One of the first signs that the attack on the leadership is complete is the open expression of intermember conflict that has been brewing since the first meeting. Once members realize they can survive conflict, and norms for its expression and resolution have been established, they move quickly to resolve their differences.

In facilitating member–member conflict, leaders ought to remember to process each interaction to completion by soliciting feelings from those involved. The intent is not to evoke a particular affect, but rather to ensure that participants are in touch with their feelings as they encounter one another.

Raising affect to a conscious level ensures that members are able to stop the process when they are feeling overwhelmed. It helps limit the possibility of members being psychologically injured.

The experience of being engaged with members who are disconnected from their feelings can be frightening. The leader might find herself taking care of them by ensuring that what she says to them is phrased in terms that she perceives they can handle. When that happens, the opportunity for mutually honest dialogue diminishes.

One remarkable event that occurs shortly after the attack on the leader is completed is the effort by group members to inquire if she is okay. This is another signal that responsibility for the group has shifted from the leader to the members. They are now inviting her to join their group as a member/leader. They now view her as more human, having relinquished some of their unrealistic expectations about her capabilities. During this transition period, members continue to seek the leader's advice. However, more and more they turn to one another for direction as members assume responsibility for the group, and in turn, for the changes they make in their own behaviors during the ascent.

Much of the leader's work is complete once conflict has been resolved. Her role in preparing a safe environment, in which members were encouraged to face, then relinquish, their dependencies, is now complete.

Members now assume many of the leadership duties. During this phase, they begin to assist in facilitating group interactions. The leader now focuses her energies on preparing the group for the Harmony and Performing stages. Although the leader is never regarded as having an equal relationship with the group members, she is accorded quasimember status, which means she is free, if she chooses, to share more personal information about herself. Now her role is that of consultant.

The conflict period in group development is very difficult to complete for a variety of reasons. Many times the group is time-limited and there are too few sessions to form the foundation needed to resolve this stage. Occasionally, the group leader is poorly trained and fails to understand the role of conflict and conflict resolution. Or, groups are continually losing or adding members so that a stable base is never established. In each case, these groups never reach the Disharmony stage and are continually characterized by expressions of indirect anger, member tardiness, and absenteeism.

The conflict period in a group's development can be compared to the adolescent period of childhood, when young people are moving to adulthood. Adolescents must eventually separate from parents and assume responsibility for their lives by adopting their own rules of behavior and developing values that reflect their own personalities. This transition period is chaotic for families.

Adolescents might begin challenging existing rules by staying out later than permitted, wearing outrageous clothing or hair styles, listening to loud and usually annoying music as a means of individuating from the family. Parents may respond to these behaviors by attempting to add more controls, which usually elicits more acting-out behavior. Many parents view the children as extensions of themselves and experience great difficulty with separation. This transition period is filled with anxiety and anger as children move toward independence and parents struggle with letting go.

There is no easy resolution to this phase. Adolescents must do battle with their parents. Their independence must be earned; it cannot be given to them. Adolescents, like group members, need predictable figures with which to do battle. Parents should understand that the conflict present during this period is necessary for the child's development and does not reflect on them as bad parents. This is also true for groups and group leaders.

The conflict stages give members the opportunity to resolve their safety and dependency needs. This period enables the group to renegotiate many of the earlier norms set by the leader. The earlier boundaries that were established by her can now be adopted by the group. Structure has been established, members feel imbued with their own power, and a sense of genuine allegiance to the group has emerged. A need for independence and expression of individuality now appears. The focus now shifts to fulfilling those needs.

Chapter 7

Showdown With the Leader: Chaos in the Conflict/ Confrontation Stage

Once the group has progressed to this stage, conflict openly surfaces. Challenges toward the leader occur. Those challenges, necessary for further group development, occur differently in groups depending on the characteristics of the leader. In the following example, this leader is challenged because he has failed to show sufficient feeling during the forming stages. In another group, the confrontation could center on the leader showing too much affect. This example highlights the difficulties facilitators face when confronted with an issue that is outside their awareness. In this example, Bill (coleader) becomes defensive and thwarts the attack. Many groups and organizations never move beyond this level of development. Without the assistance of his cofacilitator, this group may never have progressed beyond this stage. Notice how the coleader, Kate, recognizes the beginning of the scapegoating process and redirects the group's attention and attack toward her coleader. Keep in mind that although certain members may lead the attack and others hold back, each member represents the group's voice. Different group members give voice to different options at bifurcation points during the encounter. This transcript is one-dimensional and doesn't capture the multiplicity of emotions that are swirling around the group. Analysis is provided throughout the transcript. It is speculative and attempts to frame the confrontation by examining possible underlying psychological issues. This example is also used to illustrate how groups move from one level of organization to another. As the group becomes more chaotic, the stretching (movement away from one stage or point of attraction) and folding (movement back toward the previous stage) become evident. The group's stretching and folding motion depicts the turbulence (energy) necessary for growth and development. Tracking the motion would yield a pattern of a strange attractor.

EXAMPLE OF A GROUP CHALLENGING ITS LEADER

Kelly: I can feel some of the same things that Mary feels. Ah—I find myself feeling a little bit afraid of Bill or intimidated or something. Maybe it's because outside of the group you're (*to Bill*) real spontaneous and outgoing, and it kind of scares me a little bit. So I was just, I feel kind of the same things Mary does, and in the group you seem—umm—a little bit distant. So I don't know. (*pause*) I don't really know what else to say. I just wish I could get to know you better and just feel more comfortable knowing you as a group leader and as a friend (*pause*).

The group confronts Bill, the coleader, on his emotional distance from the group. The group feelings are intensified because Bill is perceived differently outside of the group. Kelly is hesitant as she begins the attack, which is typical member behavior at the start of this stage.

The group is beginning to enter the critical apex of the arc. Progression beyond this point requires that the group members tolerate upset, conflict, and often confusion. Good leadership is necessary to help members navigate this critical period in group development. The group is now moving into a period of significant instability (stretching). As it moves further from equilibrium, different choices or bifurcation points begin to appear that offer group members future avenues of development. Skilled facilitation is necessary to contain the group's anxiety until it can resolve the conflict with the leader(s).

Kate: (*moves quickly to facilitate the interaction between the group member and her coleader*) It sounds like you're talking about some of what Mary was saying, too, that if you could get him emotionally connected to you that would feel better.

Kate, as coleader, moves to assist the group in focusing the attack. Her assistance can also strengthen the group's resolve to continue.

Kelly: Yeah. (*pause*) Maybe it's my own defenses or something too that blocks up the feelings. I don't want to say that you're (*to Bill*) doing it or it's just a matter of time to get closer, but I do have these feelings. I know, I think with time they'll fade away. I hope they do.

Kelly retreats. It is difficult for her to confront the leader. She even attempts to soften or take back her observations by claiming them as her own.
Movement is uneven as the group vacillates between its current equilibrium state and a new, yet undetermined one. Pressure to accept any resolution to the ensuing anxiety is greater during this period in the group's development than at any other time.

Kate: Would it be more comfortable if they did sort of fade away, and you didn't have a want or need from him?

Kelly: (*nodding*) Yeah. I don't like to be expecting you (*to Bill*) to be a certain way, for me, because you should just be yourself—and, so that's why I just—I don't want to tell him (*to Kate*) that it's anything, that you (*to Bill*) do anything. It's more myself.

Kelly continues her retreat, even rescuing Bill.

Bill: I also feel my part in that, and feel that your feedback is accurate. I think that groups, this group is stressful for me and being in a facilitator role offers me an opportunity to hide more that I might otherwise . . . and that is comfortable, because it's difficult for me to show affect at times, particularly on demand. When that happens it's a reaction that causes me to stop and I get real cognitive . . . and kind of defensive and protective of myself.

While Bill acknowledges her feedback he also deflects it. He can see that he is defensive, but at this point can only comment on it. He also is telling the group that he doesn't want to be confronted ("particularly on demand"). He might be expressing a fear of emotional intimacy. His response is entirely cognitive.
The initial group movement generated by Kelly's challenge to Bill has almost been negated by this point, returning the group to its previously stable position (folding). However, the system has signaled its capacity for movement into possible new areas of development. Kate is attuned to this subtle signal and has responded by inviting the group to move forward.

Kelly: I can understand. Showing your feelings on demand. They're not true. So I don't want to demand anything from you.

Bill: At the same time I'm also aware that I demand that of you. As I do of each of you in the group: that you are congruent; that you attach affect to your statements; and, feel, too. So I'm aware that perhaps I'm not willing to do it. So I think your feedback on that point is very accurate . . .

Bill pushes the group back again.

Kate enlists the aid of the strongest female member of the group by asking Ann how she feels. This is a choice point for Kate. She might have chosen here to help the group by attending to Bill's lack of affect, helping him get in touch with his feelings. However, Kate may also be hesitant to take Bill on directly, too.

This is one of several bifurcation points that begin to appear in this group. At some point, movement beyond one of these critical junctures will ensure that the group cannot return to its previous state. However, at this point, the group is at the precipice of the vortex, still quite capable of retreating. However, critical mass is slowly building as Ann now joins Kelly in the confrontation.

Ann: (*Identifies the essence of this attack. Bill is disconnected from his role as leader.*) Well, I can relate to Kelly's statement. I feel that in the group you're (*to Bill*) the one we know the least, and that makes me feel bad. It seems like that the primary thing that's happening is the facilitator role is here and the person really isn't. I guess that's what bothers me a lot. Because I feel that everyone of us is kind of up front and revealing real emotion. And, I feel that your (*to Bill*) emotions are kind of within. So that's what's causing me some distress today.

Ann picks up the attack.
The group is perturbed again.

Kate: (*to Ann*) What is it that you want from Bill today? (*Kate encourages Ann and focuses the confrontation.*)

Ann: (*cautious, and avoiding direct confrontation by demanding Bill show affect*) I guess reaction on a feeling level. Not to me particularly. I don't at this point think I need a reaction on a feeling level, but for those who need it. I know it just doesn't come naturally, but that's the frustration. I feel ... If you (*Bill*) were totally relaxed in our group than you would just be natural. . . . I feel that you're not comfortable with us. I'd like everybody to be comfortable, that's my bias, that's my problem.

Ann makes an indirect request for a feeling by asserting that others may need it from him.

Kate: So you want to know if he's comfortable here, or how he could get comfortable.

Ann: Well, feeling I think. I think if you (*Bill*) felt a part of us that you'd just be more than the facilitator in charge of us, be yourself.

Mary: . . . what a conflict that is.

Ann: (*nods*) Yeah.

Mary: (*to Bill*) I can appreciate where you're at right now.

Ann: (*nods*) Yep!

Mary: . . . I'm not saying it well, but I think there is a fine line there. 'Cause personally I want to keep Bill the facilitator and I think I can respect his wish to, to at times or whenever draw back on affective things. . . . I can respect that. (*To Bill*) Even though I want a feeling response from you. I was okay with you saying it just wasn't there. I mean all of a sudden I understood. I didn't take that personal. I found, at first, I was taking that so personal—a—I was okay with that. I really was. I mean I respect that because there are times that you (*to group*) do get to put your affect aside or something for awhile. And really you know every one of us can do it. Look what Doris did for me. She gave me a hug, that's affect. She gave it to me so it's not like Bill has to do everything.

Mary follows with what is probably a realization, albeit unconscious, that dethroning the leader, seeing him as human, and giving up the perceptions of him as omnipotent means that group members must take charge. However, if you keep the leader then you relinquish your power. During the Conflict/Confrontation stage, members vacillate between dependence and independence. Gender expectations might be at play here. Mary might be willing to allow the male leader to withhold his feelings although females have been expected to share. The group partially recognizes that they can lead and attend to one another themselves. Overall they are rationalizing the leader's behavior and avoiding anger.

Another critical choice point appears. Movement beyond this point would plunge the group into the vortex. This point is an example of an edge of complexity or stillpoint in a system. The group teeters between one state and the possibility of another while avoiding total chaos. A very small push by the leader or another member at this critical point could have dramatic effects. Those leaders with very keen perception and good intuition (nuance) would sense this opportunity.

Kate: You sound pleased to me that all your hopes don't have to lie with one person, that it can get spread out in here.

Mary: It is a relief. Because what I found myself doing was in a conflict—putting it all on Bill or you and feeling all this responsibility in myself for some crazy reason. . . . I see everyone having parts—so much to offer—and as a group we all have our part. I feel really good about this group.

Mary deflects the anger and distracts the group, urging them to stay away from it ("feel good").

Group: (*silence*)

Jill refocuses the attack and escalates it.

Jill: (*to Bill*) . . . But I guess I need to say something else, too. And I feel bad about it (*crying*) because the exchange between Mary and Bill was real good for you (*to Mary*). (*To Bill*) I'm not sure about you, because you don't show it. What Kelly said and Ann said I feel really bad. They've given you permission to behave like that. I feel that they should, and we should all let people be here how they want to be. Yet, I'm really angry at you for that because you're expecting us to give of ourselves and I don't feel that you are. So I'm angry at you.

Her point that he expects them to act in a way that he is not willing to is the crux of much of the dissatisfaction and anger with group leaders. However, their early role as the group's protectors does not allow them to show much vulnerability. Hence they inevitably find themselves caught in this dilemma. Jill directly expresses her anger!

The group is moving more rapidly between stability and possible change as you can observe from the short time frame between their retreat (Mary's previous statement) and advance (Jill's statement). This increased fluctuation provides the necessary energy to propel the group forward. Kelly, Ann, and Jill now comprise the growing critical mass necessary to propel the group forward.

Kate: You sound angry, like you're not getting enough emotion from him personally here.

The coleader focuses on Jill's anger as a means of directing the attack. Again she is at a bifurcation point. Bill has been silent and has been indirectly and directly confronted. Kate could attend to Bill here and get him involved, feeling, and back in the group. But she chooses to encourage Jill, which might mean she is avoiding Bill or abetting his withdrawal or trying to empower the group.

Jill: (*hesitant and backing down*) But I understand that that's him and (*to Bill*) I understand that that's you, and if you don't want to do that, I guess I don't think that you should have to. You know so I see both sides of the coin there. . . . I guess I want to know you. It bothers me that we don't see that here, and I don't know. I am struggling with that. I understand it, yet I don't like it. And that's selfish of me I guess and I don't know. I'm feeling lots of different things. . . .

Another example of the uneven movement during the attack. Members move forward, then retreat. The leader's reactions are carefully monitored by the group members.
The group is stretching and folding.

Kate: Because so much has been stirred up in you?

Jill: (*crying, nods*)

Jill's emotional response to this confrontation indicates the powerful emotions that are evoked when confronting the group leader, particularly when the leader remains distant and unable to respond empathetically.

Mary: I feel that too, really lots. . . . I think what you (*Jill*) did today is really—you took a risk.

Jill: . . . It feels better to get it out if nothing else.

Mary: I'm just starting to realize how capable everyone is here. You know I think I was doing a lot of underestimating of people and myself and all of a sudden I realize that.

Mary diverts the attention away from the leader to other members to reduce group anxiety. The attack can be directed and focused, but it cannot be forced. Members with proper guidance will move at a pace commensurate with their feelings of safety. The coleader's role is to continually invite them to move forward. As with many of the interactions that take place in groups, the spokesperson, in this case Mary, speaks both for herself at the manifest level and for the group at the latent level. Her hesitancy reflects the group-as-a-whole's caution.

Jill: . . . I really feel better having talked. . . . I feel more in control. I can talk again. I feel better.

Jill is relieved that the attention has been shifted from her confrontation with the leader. The group moves on to discuss the lack of affect from other group members, and after about 40 minutes returns to Bill, who has sat quietly during this period. The group moves to invite him to join them. On the manifest level, they are asking him to participate in the group. On the latent level, they are asking him to reveal himself and join them as equals.

Jason: (*to Bill*) I'm wondering if you're alright?

A male member now requests Bill's involvement.

Bill: Yeah. I've felt excluded for the last 40 minutes and not really able to participate or speak—I've been so aware of the time and the lights. I have felt very much out of the group, disconcerted . . .

Bill refers to the videotaping, which may explain part of his reluctance to reveal himself. He continues avoidance behavior.

Kate: You need anything?

Bill: No, thanks for asking

Jill: Are you angry with me?

Bill: (*controlled*) No, I think your feedback was accurate, appropriate.

Mary: Do you feel like I'm pushing you out again. I feel like I did. . . .

Ann: (*to Bill*) . . . I want Bill to know I really do like you. I don't want you to construe my feelings as being negative toward you.

The group is inviting him back. He remains distant. Jill wants to make sure she has not displeased the "parent." Mary, too, wants approval for her actions. Ann joins them. Again you witness the tremendous power that is projected onto this leader. His withdrawal and silence increases the group's anxiety and manipulates the group to his side.

Group stability remains fragile. A small nudge at this point could send the group tumbling into chaos.

Bill: I didn't. Thank you for that.

Jason: It's almost time to go. Bill said he felt outside the group. (*to Bill*) Are you still feeling that way?

Bill: Yeah. I'm feeling a little outside of the group.

Bill is almost pouting here because the "children" have challenged him. Keep in mind that a blind spot in Bill has been activated and he cannot see what effect his behavior is having on the group. Kate's help is needed here. Jason has temporarily assumed the leader's role, even to the extent that he is monitoring the time boundary.

Jason: Anything we can do?

Bill: (*with no feeling*) . . . Thank you for asking, Jason. I feel I have an invitation to come back when I want to come back.

Jason: Okay.

Jill: (*irritated*) Does that mean you're choosing to be outside of the group today?

Jill's anger rises and she confronts Bill. His unwillingness to rejoin the group allows him to retain power.

Bill: Well, no. I don't think I was choosing to be outside of the group, and I feel I've had several invitations now to come back. . . .

Mary: I want you to return very much.

Mary is firmly ensconced in the placater role. Remember, she also reflects the group's hesitancy to confront Bill.

Kelly: I do, too. That's probably why I mean, all of us have expressed our feelings toward you because we want to get to know you.

The session ends with the facilitator still outside the group. Anxiety remains high because no resolution of the conflict has occurred. Shortly after the next session begins, the coleader confronts Bill.

Kate: (*exasperated*) . . . I have felt—I thought a lot about the frustration people have expressed in here during the last session about not sensing emotion from you and I'll speak for myself at this point. . . . I have felt disappointed, angry at you at times for the lack of that.

Between sessions Kate has identified her own feelings toward Bill and expresses them. She realizes, too, that Bill's distancing has neutralized the group in their attack. Her confrontation serves as a model for the group. She has also assumed the role of group spokesperson.

Kate's intervention adds to the earlier mass created by Kelly, Jill, and Ann and provides the necessary energy that pushes the group into the vortex of transformation. There is no certainty how the group will emerge from this turbulence, other then it cannot return to its previous state.

Bill: (*calmly*) You would have liked more of that from me.

He deflects her anger by trying to facilitate her interaction and reassume control.

Kate: (*angry*) And I was thinking about your statement of emotions on demand that you couldn't give that. And I didn't hear that from people . . . that doesn't mean that wasn't there. I should just say I didn't want that from you. I didn't want emotion on demand, but what I did want, I realized was some vulnerable emotion, be that anger or hurt, but something where you stretched as I and others did here. I didn't feel that you emotionally stretched here. . . . I'm wondering how you are feeling about that?

Kate is angry and the attack is now direct!

Bill: (*agitated, and stunned by the confrontation from his coleader*) . . . I feel blindsided by that. Ah—and while it may be accurate it doesn't feel real good to hear that from you.

Bill is caught off-guard. His authority is questioned by his coleader. He feels betrayed, as if it was her responsibility to protect him from having to reveal himself.

Kate: You look like you feel hurt.

Kate presses for an emotional response. Remember, to this point the group has no idea what Bill is feeling. They can imagine it to be rage, which keeps them distant. Bill appears hurt and attempts to counterattack Kate by chiding her for not confronting him yesterday.

The group is very unstable at this point. Both facilitators are engaged in the conflict. But their earlier work in the previous stages has created the necessary norms and boundaries to contain the group during this period.

Bill: (*teary-eyed*) I do. I feel hurt, I feel blindsided by that. I don't know what you want from me in regards to that. But ah—ah—but it didn't feel clean to me ah—maybe because it didn't happen in the moment when it was occurring, it's happening now and just doesn't feel clean to me.

In another effort to divert the attack and regain his composure, Bill turns the confrontation back on Kate.

Kate: (*reasserting herself*) . . . I'm sorry about that. I can appreciate where you feel blindsided and it wasn't until I had a day to think about it. I won't excuse myself for it. . . . It's the best I can do, now.

Kate won't be deterred. She pursues Bill.

Bill: (*nods*)

What happens next is a clear example of how group members facilitate interactions, especially conflict between coleaders, when proper norms and boundaries have been successfully negotiated during earlier stages. Jason leads again.

Jason: We got to clear this up, the feelings between the two of you, right now. (*to Bill*) Are you okay with her right now?

Bill: (*defensive*) Yeah, I said what I needed to say. The feedback was heard and noted. (*to Kate*) I don't object to what you said. I don't find anything wrong with the feedback at all. Ah—just how I felt about receiving it at this time.

Jason: (*to Kate*) Why the tears, are you okay? Or is there something else?

Kate: It's like a post reaction. It was hard for me to do . . . I felt sad. I needed to do that. That's where the tears came from.

Kate reacts to the stress from the attack.

Jason: Feel better now?

Kate: Probably in about 5 minutes I'll feel better. Thanks for attending to me.

Bill: (*calmly to Kate*) Do you need any more from me?

Kate: No not a bit. In fact I was real clear I didn't need anything from you except I needed to tell you that. . . .

The attack had been thwarted. Bill remains emotionally distant. Kate has been pushed back. The group is blocked, uncomfortable because the conflict has not been resolved. The group will begin to resort to destructive or regressive actions if the conflict remains unresolved.

Group: (*silence*)

During this stage, or any stage for that matter, if the facilitator remains blocked and is unable to respond emotionally or in some other way the group needs, members will begin to take responsibility for the impasse, much as children do when a simmering conflict between parents begins to heat up. They will act in ways to refocus the attention on them creating a kind of double-bind situation. Because children or group members are in a one-down position, unable to comment on the leader or parent's stifling behavior, as a result scapegoating can occur.

The group will remain unstable while it searches for solutions to the impasse. Containment of the immense anxiety experienced by group members cannot be over emphasized. The group will self-organize if properly contained. Unfortunately, most groups who reach this critical point fail to contain themselves long enough for a successful resolution to emerge. Short-circuiting this stage then limits the group's creative potential found on the other side of the arc.

Doris: I'm feeling responsible for all the attacks Bill has had to sustain. I'm not sure how to put it into words, but as I was thinking over the group. . . . I saw myself setting up a negative norm early in the group when I cried because of the attack I sustained from Jason. (*to group*) I think because of that you have hesitated to share any kind of negative feelings toward each other and you keep testing Bill because he seems like the only one that can handle them. I don't know how to reverse that except that I want you to know that I'm ready for any kind of negative thoughts about me. I can handle them now. I just wanted you to know that.

Doris "volunteers" for the role of scapegoat. She diverts the group's attention onto herself and claims responsibility for the turmoil. Her behavior might reflect her desire to deflect attention from the parents' (Bill and Kate) fight. This may be a role she occupies in her own family. Doris's intervention offers one possible solution for the group. Although it is regressive, many groups not properly prepared for this stage settle for any resolution to the anxiety, even a destructive one. Although no outcome can be predicted, successful resolution of the major issues of concern (safety and affiliation) in the previous stages maximizes trust and establishes enough group cohesiveness to help the group successfully complete this stage.

Kelly: That was certainly a big risk.

The group's focus shifts toward Doris as she asks for feedback about her group behavior. Members respond. Bill keeps the focus on Doris.

Bill: (*strongly*) I'd like to give you (*to Doris*) some feedback and (*to Kate, almost as if requesting her collusion*) I'll need your help with this.

Bill is telling Kate to get back in her role as coleader. However, Kate recognizes that the group is moving in a destructive direction. She offers them another "more correct" solution by refocusing the energy and conflict toward Bill. This is one more example of how a very small and properly timed intervention can constructively move the group.

Kate: (*redirects the focus*) Alright. Before you do that I need to say how I feel—the only thing I've been wondering—when you talk there is a lot of energy and the only thing I've been wondering if you're angry at me and if you are I'd like to hear it.

Bill: (*curtly*) No, I'm not! I think I'm angry at what happened, but at the same time pleased that you took the risk.

Kate: (*helping him to focus on a feeling*) So you're somewhat mad that I wasn't—you know.

Bill: That you don't do what I want.

Kate: (*laugh*) Like my husband.

The marital metaphor surfaces, which may provide clues to the underlying dynamics of Bill and Kate's relationship.

Bill: That's what I'm angry about is that you don't do what I want.

Bill seeks collaboration from Kate, not equality.

Kate: When you want it, and that's what I was . . .

Bill: (*interrupting, loudly*) You're my cofacilitator and I expect you to cover my back and do what I want. And when you don't do that I get angry. And at the same time after allowing for that anger to be there I realize you're a person, too. You know, because you work with me doesn't mean you have to sell out your soul, and ah—ah—do what I want. So, yeah, I'm angry because you don't do what I want.

Bill is finally connected with his feelings. He expresses anger, which still maintains his emotional distance. He pushes Kate back, asserting, now properly, his leadership role. This is the point where leadership is earned by the group, not given away by the leader.

Kate: And I'd like to say I was mad at myself for not being quicker—I spent the night—

Bill: (*interrupting, with force*) It's hard for you to be quicker when you get caught up in trying to do what I want.

Kate: (*with energy*) Yes it is. That's right and that's where I was mad at myself . . . and then I thought if I don't come back and tell you what I was trying to do—what I thought you wanted me to do and—ah—mince my words then I'll never get off of it. And I thought as long as you and I end up working together—if at all after this—it's going to keep coming up so I thought this is the cleanest way to do it. But I want you to know that while I was mad at you I also had anger at me because I wasn't quick. So I can see where you're mad at me.

Bill: And also I think cleaning this up is caused—somehow we didn't— that I didn't, that I didn't get it all out with you. That I was angry 'cause you weren't doing what I wanted and that seems to me has affected the group and what's going on now. That's the other part of it.

Kate: Well, thanks for saying that. Yeah, I think . . . I'm glad you expressed your anger and said it to me.

*Finally demonstrative, Bill reveals himself to the group. He speaks to Doris almost
as an apology for allowing her to take responsibility for his behavior.*

Bill: (*openly with emotion*) (*to Doris*) . . . a large part of how I see myself—that my emotions are not always readily accessible—is that
I'm very defended in terms of how I feel about things. I'm always
protecting myself against what I might hear that's uncomfortable
or unpleasant. When that happens I shut down. It's an automatic
thing with me. I shut down, and respond in a very calculated
way, very disconnected from my feelings. So when I give feedback
to people that I see as similar to that I'm also aware that it could
be a lot of my own stuff that I'm projecting over there . . . as a
result of confrontation I know what happens to me. When people
confront me . . . it's automatic, it's that instantly I freeze up. I
stop breathing and I get disconnected from my feelings and then
all I can do is think. Then when people say "how are you feeling?"
it's like I'm not even breathing. I don't know what I'm feeling
I've just shut down . . . it affects my relationships now. The feedback I've heard from you and other people has been very accurate. . . .

*He has revealed his struggle, exposed his vulnerable side, and emoted in the group.
Bill shows his humanness. He speaks to Doris as a way to take back his role in
attempting to scapegoat her. The members move immediately to acknowledge his
humanness.*

THE GROUP BEGINS THE DISHARMONY STAGE HERE!
The group affirms Bill and accepts him as the person he has revealed himself to be.

Doris: I just have so much respect for you right now because you've
been able to admit you're not perfect. It makes me feel so good
toward you.

Jill: I really want to thank you for that, too, because I was so angry
with you sitting back and it seemed like you didn't care at all.
That's how I felt. And today you just told me something. I've
learned more about you in the last 5 minutes than since I have
known you. You know I really thank you for that. . . . I guess
that's what I was looking for. It meant a lot.

Bill: Thanks. I don't know, it's always the stuff that I hold onto, that
I give so much weight to, that really when I let go it seems like
. . . there was really nothing to it.

Kate: I feel really warm toward you in what you said. I have a lot of respect for you for opening yourself up like you did.

Bill: I think part of the reason I push the way I do, and push each one of you in here is a way of wanting to push myself more, wanting to demand that I show more emotions, that I be more connected. . . .

Bill and Jason talk about the similarities between them.

Kelly: (*to Bill*) . . . I see the good side of you that I've never seen before and also when you talked with Jason I learned a lot about Jason that I didn't really know before either. I'm just glad I got to know that.

In this transcript, members are shown challenging Bill on his emotional distance from the group. It is the women who undertake this task, who value emotional intimacy. Although the group attacks the leader on his lack of affect, another issue at the latent level may exist. The cultural influence of gender roles and their expressivity is at play in this group. Bill, a skilled leader, under stress from the group to reveal himself, regresses to the traditional male role, unemotional, unexpressive. His cofacilitator, Kate, may have taken on the unconscious role of wife. Bill expects her unquestioning support and feels betrayed when she challenges him and expresses anger at his emotional distance. Kate, here, is at some level aware of Bill's projection onto her and feels manipulated by it. She is hesitant to do more than elicit or clarify the feelings of the other women in the early part of the transcript. She does not really support them by validating their feelings for the risk they are taking in confronting Bill. As a result, most of them are reluctant to push Bill too hard. Kate avoids drawing Bill into conflict until the following session. It appears that the women, including Kate, are struggling with their cultural roles. The issue of emotional intimacy is a common one that women raise with men in their lives. So the intensity of their feelings toward Bill, and possibly other men in the group, may reflect their frustration over this issue with their husbands or fathers. What is so threatening about the conflict is that it may lead to a break in the relationship and result in abandonment, just as owning and sharing emotions seems threatening to Bill. Because of the cultural conditioning of the group members, they demonstrated an impressive amount of courage and honesty in acknowledging and confronting these issues. This is surely the source of the generative potential in this group. By the end of the transcript, the conflict appears resolved to the satisfaction of everyone. In fact, some role reversal took place. Bill made himself vulnerable, human (feminine); Kate stood her ground (masculine). The group sees the leaders as human and accepts that, paving the way for their own assumption of responsibility for the group.

This interpretation is one of many that can be constituted to explain the actions of this group. There are many completing and compelling alternatives. No one explanation is correct. It is offered to raise your awareness again to the many opportunities for learning that can take place in groups. Once the conflict is resolved, the leader might decide to share this interpretation of events with the group members. Exploration of gender roles and their influence on these group events might prove instructive for the members.

Although the transcript is one-dimensional, it does demonstrate how critical mass is achieved and group change occurs. Uneven group movement evidenced by the stretching and folding that occurred is characteristic of how groups evolve. Skilled leadership was also illustrated by Kate. Her recognition and intervention at critical moments helped guide the group. During the turbulence small, correctly timed interventions by the leader or any group member can have a dramatic impact. Once a group enters the vortex, no outcome is guaranteed. The eventual solution the group chooses is dependent on their history and could ultimately be destructive for them if they settle for an unhealthy, but quick resolution to the overriding anxiety they are experiencing. Groups or organizations that are led by authoritarian leaders, who are controlling and withhold information, are doomed to failure and operate well below their full creative potential.

DISHARMONY

Once the group has taken power, members sometimes conclude that their work is finished. After all, they have just successfully navigated the turmoil of the Conflict/Confrontation stage. The leader's role now is to keep the group from becoming too content. Primarily, her role on this side of the arc is to perturb.

The group now sees the leader as more real, having shed some of their misconceptions about her. Members can be very solicitous toward the leader during this period, so it is very tempting to stay longer than necessary. It feels good, and if the leader has gotten the group to this stage she has earned it. But this is really only the beginning. The real potential for therapeutic work lies ahead, so the group must be pushed forward. Furthermore, it is time for the leader to relinquish some of the leadership functions to the group. To accomplish this, she should be hesitant with her guidance, giving the group time to make their own interventions. The leader's role can be limited to process commentary while the members assume responsibility for task and maintenance functions. Members should be encouraged to attend to one another (e.g., asking Mary, "Are you curious how Ellen is feeling right now? Why not ask her?")

Members should take charge of time boundaries, starting and ending the group on time. The leader's role moves from the foreground to the background, nudging when necessary. Group members consult the leader when they need assistance. During this transition period, members may be caught between their lingering dependency on the leader and their newly earned freedom. Simple gestures like avoiding direct eye contact with members or using hand signals to redirect communication patterns are helpful in reducing the leader's central position. She should practice being invisible.

There may still be unresolved conflict among members. This must be addressed before the group is fully into the Harmony stage. This does not mean that conflict does not occur after this stage, but anger and animosities generated in the forming stage must be addressed here. All members, especially those still at the margins, should be fully included in the group. This might require additional disclosures, confrontations, or clarifications among members.

During the Disharmony stage, members act to include marginal members. The following transcript depicts a fairly common group experience. A group member who has been present at all meetings but has not actively participated is questioned by the group. Group members are able to recognize some level of involvement and commitment on her part, but view her silence as a distancing mechanism. The chaotic activity of the forming and conflict stages allowed her to be carried along. But now marginal members are expected to participate fully. It is the members and not the leader who facilitate this interaction, a distinguishing characteristic of this stage.

Jack: ... I'm sensing a frustration, Helen, that maybe you haven't given as much to the rest of the group as what you've wanted to when you first started.

Helen: (*firmly*) You see I think I have, but I don't think that people have perceived that I have. But I have emotionally. I felt more, parts of what's been going on. ... I think I've had more feeling reactions, but I don't think the group has seen that, and that, I feel bad about ... the one thing I've learned this week ... is that you don't always have to say anything in order to be effective. I think if, if we're really communicating people can see when there's an understanding, and reaction, but I keep thinking everybody knows me cause I'm as open as a book (*laughs*). If I'm mad people know, if I'm sad people know, I don't keep anything in.

Barb: You know though, Helen you are—ah—when you say that I know what you mean. Because just in social situations outside you're very verbal and outgoing, but you haven't been that way at all in group. So in group I don't perceive you at all like an open book. I perceive you more as a real mystery.

Helen: I feel like if there's nothing to say that contributes something about my feelings—just to speak—I think that's a waste of time . . . . I don't feel that I've been holding back, but maybe that's a defense mechanism. I don't know. I thought about that ever since I was confronted with that by you (*Barb*) maybe I have been holding back.

Barb: . . . I've missed out on getting to know you a little better in group. Although, I've really respected how you could sit back and listen. And I really believe you when you say you've taken in a lot of things from the group.

Helen: (*interrupting*) But not shared.

Barb: Yeah, I guess so. . . . I didn't get to get something from you because you were quiet. So there's some disappointment there. Maybe I'm expecting too much from you.

The group continues giving Helen feedback on her participation in the group. Her obvious emotional involvement with the group had permitted her to be carried along to this stage. Now the group is requesting more.

Shelley: I have really enjoyed your presence, Helen. I have noticed the affect, except I would have loved to hear more specific explanations of what the affect was about. I can't read into your mind, what you're maybe tearing up about, or what you feel happy about. I don't know the specifics, but I can see the overall thing, somehow it makes me feel secure. Just that you're there, because I feel you're feeling. I need definition for it, more definition.

Robert: How are you feeling with this feedback?

Helen: I'm feeling frustrated (*smiles*).

Robert: The feedback seems incongruent with how you see yourself in this group.

Helen: It is terribly incongruent to me. I think there's ways of sharing that don't involve spoken communication. And my interpretation of how I feel about the group and the sharing—I think I've shared my feelings.

Shelley: I feel uncomfortable with it. If you don't explain to me what you're feeling I might misunderstand. I might read things into it that aren't exactly right.

Helen: But see that's where I think the group hasn't heard me, because when you were hurting that day I related to that. . . . I felt it when Barb risked . . . I said I really didn't do my job, I feel guilty. I think I was risking when I said that. That was a risk for me. That was a great deal to even say I felt guilty, because I had held back. And (*to Shelley*) relating to you, I hurt for you that day, but I sat back, but at least I acknowledged it.

Shelley: You know what I wanted from you right back then. I remember it clearly. I wished you would have given me more and said "I have a lot of empathy for you because" and then shared what it was about yourself that specifically I could tie into so that I could understand the empathy better.

Barb: . . . 'cause when I peek over at you now and then I can see that you are really involved. I can see emotion going on in there. How much you care about what's going on in there . . . but that's it. You're assuming that everyone can see that so therefore they know you better, but I don't think we do see that because usually we're all focused in, just as you are on wherever the interaction is, and we miss that unless you speak up and tell us . . . that's what I would like.

Helen opens up and shares how difficult it is for her to be the focus of group. She is reluctant to be the center of attention.

HARMONY

In the Harmony stage, members may become restless, bored, and sometimes discontent with the group if they are not working on the issues of independence, intimacy, and risk-taking. The group's development is not predicated on them doing so. Unlike the urgency associated with resolving safety, affiliation, and belonging issues, group members feel safe and content with the level of the group's development.

Profound and lasting change can be accomplished during the ascent. Asserting independence within the group, disclosing intimacy needs, and taking risk through self-disclosure allows members to explore themselves at deeper levels. Now the group is free to focus on members with greater objectivity because personal relationships have been established. One way this is accomplished is through acknowledgment and acceptance of one's dark side, that aspect of the personality that members kept hidden from others. Essentially, this opportunity permits members to disclose and integrate fragmented parts of self. At this stage of group development, the opportunity exists to reveal oneself fully and experience the acceptance of others. Often it is remarkable to witness group members reveal some aspect of self that they hid from others, only to find that the disclosure draws people to them, not pushes them away. These group self-disclosures facilitate self-acceptance. During this phrase of group work, significant therapeutic change can take place. The change is akin to restructuring self-perception, rather than simply altering behavior.

The group, now, is responsible for its own direction. Leadership is shared, cooperative. Each group member has demonstrated some area of leadership ability and maintains responsibility for it. The leader is now only one source of expertise. However, there are some areas to which the leader must still attend.

She should continue to push the group forward, by encouraging self-exploration. She should help members clarify areas in which they want to do personal work. Her key role during this period, however, is to monitor the level of group cohesiveness.

Remember during the forming stage when the leader had to contain the group's anxiety to insure that members were not overwhelmed? Now she must balance the cohesion in the group between too much collaboration and too little tension. The leader now works to increase tension when necessary.

A critical source of personal learning that takes place during the ascent is the result of feedback from one member to another. Too much cohesion among members inhibits feedback, too little makes it unsafe. The leader helps maintain a healthy balance. Giving and receiving feedback is a primary source of therapeutic growth in groups.

FEEDBACK

Feedback is a process whereby members share their experiences of one another. It is another unique feature that distinguishes group work from other forms of therapy. In the group milieu, members have a rare opportunity to find out how their actions impact others. Furthermore, they are afforded opportunities to reciprocate.

Feedback is an important aspect of group development in the ascent. Once a group has established a climate of trust, feedback no longer needs to be couched in cloudy, ambiguous language. Here are some steps the leader can follow in teaching group members how to give and receive accurate feedback:

1. *Create a receptive environment.* Members should be asked if they want feedback. Ensure that the member has latitude to say no.

2. *The feedback should be specific and directed toward a behavior that the recipient can do something about.* Telling a group member you don't like the color of his or her hair is hardly productive.

3. *The observed behavior should be drawn from actions that have occurred in group.* The more examples, the easier it is for the recipient to understand the feedback.

4. *Feedback should be phrased in a language that can be heard.* Blaming statements should be avoided. The use of "I" statements rather than "You make me feel . . ." statements creates acceptable feedback. In addition, ownership of the statement is maintained by the sender.

5. *Share impact.* The member giving the feedback should also share how the actions of the recipient impact him or her. For example, if one group member gives feedback to another by telling her that her loud voice gives him the impression that she is angry, he should complete the feedback by sharing how that behavior affects him. In this case he might say, "I feel intimidated by your loud voice and I find myself avoiding contact with you." This step validates the feedback. It lets the recipient know what impact the specified behavior has on another person.

6. *Process feelings.* Once the feedback is given, it is necessary that the recipient be given the opportunity to respond, particularly with a feelings statement. The feelings statement lets the group know that the feedback has been received. Occasionally the member receiving the feedback gets defensive and is unable to immediately experience its impact. The leader's task is to help the member connect with his feelings. On occasion members leave the group after having gotten feedback and felt its impact, but being left to process it alone. The group must be slowed down so that this important step is not overlooked.

7. *Get other opinions.* If the feedback seems unwarranted or unfair, or if the recipient wants to hear from other members, have him ask the group. Remember to ensure that each interaction gets fully processed. If each interaction is processed to completion, emotional overloading can be avoided.

8. *The leader should avoid giving feedback first.* Other members should be encouraged to begin or take the lead, because the leader's feedback can carry more weight. Frequently, members turn to the leader when something difficult needs to be said to get the group moving. Group members are to be supported in taking the initiative.

Here's one example of how feedback can be initiated.

Betty: I'd be willing, if anyone feels any, to take some feedback. . . . I've heard some very nice things about myself from many of you . . . and that felt good . . . but if anyone has some feelings they previously felt uncomfortable saying and want to say those now I feel really okay about it being directed to me. Because it would be helpful to me. . . . I'm not perfect and if there are some other things people notice . . . I want to know. I think I've identified for myself what some of those are, but I'm interested if anyone else sees some things that they find are obstacles for me in relating to other people.

Betty asks Gina for feedback.

Gina: . . . I was going to say something, trying to put words to the obstacle I felt with you and there's no fancy way to say it. Your voice is very soft, that's how I hear it, and that's an obstacle for me in communicating with you. It's hard for me to hear you sometimes. It's frustrating for me to ask you to speak up and mostly it's an urge. What I want to say to you is "So what, just own your strength and get out there." You know (*slaps hands together*) that's like what I want to say. I've been really impressed in here with your strength and that's the only obstacle I felt . . . it doesn't seem like an obstacle, but I felt wanting, that's a better way to say it. I've wanted you to sort of rise up at times and your voice is the way I heard you being quiet.

Betty: And, I can recognize that theme from day one. I was the one no one could hear . . . yep, I can see that, because I do often feel more, or a strength and conviction about some things, but it's difficult for me to express it that way. Sometimes I say it, I do say it with such a quiet little voice that I suppose it's not very congruent. . . .

Gina: . . . Sometimes I fantasize, and I am now. Gee, Betty you have definite views and perceptions on things and I think you would be a very powerful woman if you'd just gave it that—you know put it into fourth gear and went for it. That's it!

Betty: (*louder*) I really hear what you're saying. It makes sense to me and that's real helpful for me. . . . I've come to the point, and I'm really glad, where I can take some negative things and see how much good that can do me to listen to that, 'cause in the past . . . that's when I really often tuned off and got such hurt feelings and said "Oh, don't criticize me and don't say anything bad." Yeah, I feel good about that. I guess that's what I was asking for today is to test that out. To take some criticism, and take it with an attitude of how it could help me to grow, rather then become upset about it.

Others offer feedback to Betty, and Sam, the coleader, urges Robert to share his reaction to Betty's quiet voice. A critical, and often forgotten, portion of the feedback cycle lets Betty know the impact of her behavior on one member.

Sam: (*coleader*) Robert, how does Betty's behavior affect you?

Robert: I think the feedback is accurate. When I hear you speak with a quiet voice I sometimes check out. In other words, it's hard for me to stay with you when your voice is soft so while you may be saying things that are powerful I find myself sometimes drifting off . . . so my part in that is sometimes I don't hear what you're saying.

Betty: Yeah. I can see that, too. Earlier in the group I remember experiencing that—that's where my feelings of isolation come from that I don't feel heard. Well, I'm hard to hear when I don't speak up—so I can understand that and I've been told that before but I never connected that up until now.

Let me add that few laboratory or training groups reach this stage of development when sessions are limited to 10 to 15 meetings. These are too few sessions to build the necessary foundation for genuinely achieving this level of development, particularly with neophyte group leaders.

During summer sessions, we run 5-day group experiences where all participants stay in lodges at the site. Group sessions are held day and night. In this implosive environment, most groups are quite capable of reaching the Harmony stage.

The final stage in the arc, performing, is somewhat of a mystery. Groups seldom get there, and very few group leaders have actually experienced this stage in groups. However, many have had similar experiences in other aspects of their lives. These have been referred to as peak experiences and represent a kind of self-transcendence: a loss of self-consciousness or self-awareness. It comes from being totally present, absorbed, fascinated with the experience at hand. It has a timeless quality to it.

As a research assistant in graduate school, I had numerous timeless experiences while working in the computer center. It was a cinder-block building with no windows. I would sometimes enter the building around noon and become so absorbed in my work that I was totally unaware of the passage of time. On numerous occasions, I left the building and found that it was 3 a.m. There was an unreal sense to it, walking outside into the night, thinking I had just been working for a couple of hours and finding that it had been 15. On rarer occasions, with my writing, I can experience a loss of consciousness by becoming so absorbed with my work. Unfortunately, the occasions are not frequent enough.

Self-transcendence is the Taoist state of acceptance, letting things happen, rather than making them happen. It is a state of nonstriving, non-wishing, noninterferring, and noncontrolling, the state of having rather than not having. It is possible for group members to achieve this state, but it is rare, although there are times when a group becomes so focused in the moment that a loss of self-awareness occurs. However, what I am talking about here is a sustained state that is achieved by the group-as-a-whole and provides transcendent experiences for the members. I have only experienced one such group.

Inclusion of this stage in the arc represents the possible fulfillment of the group's potential. It is important to acknowledge that although most groups never reach this stage, they catch glimpses of it throughout their development. Leaders should recognize the possibilities inherent in this stage and hold fast to a vision of groups that incorporates the transpersonal realm.

Author's Note: Some of the material in this chapter previously appeared in B. A. McClure, *Small Group Behavior* (Vol. 18[2], pp. 179–187). Copyright © 1987 by Sage Publications. Some material appeared in B. A. McClure, *Small Group Behavior* (Vol. 20[4], pp. 449–458). Copyright © 1989 by Sage Publications. Reprinted by permission of Sage Publications, Inc.

Some material in this chapter previously appeared in B. A. McClure, *Journal for Specialists in Group Work* (Vol. 14, pp. 239–242). Copyright © 1989 by the American Counseling Association. Reprinted by permission of the American Counseling Association.

Group Metaphors
as Strange Attractors

The group process, whereby members are able to alleviate anxieties and fears by indirectly addressing issues of concern that are too risky or threatening to openly address, has been called by many names: the group "theme,"[1] the "group mentality,"[2] the "group fantasy,"[3] and, more recently, the "group metaphor."[4] *Group metaphors* represent words, analogies, non-verbal expressions, and stories in which "thoughts and feelings about an emotionally charged situation have been transferred to an analogical situation that preserves the original dynamics."[5] Similarly, *metaphors* are defined as analogies that permit group members to substitute "a nonthreatening external subject for a threatening internal one, enabling them to experience affectively charged worlds of meaning from a safe distance."[6] A primary function of group metaphors is to provide relief from excessive anxiety. Metaphoric language shifts the group's focus from the manifest or conscious level to the latent or unconscious level where the group can work through shared problems and anxieties. This movement from one level to another enables the group to remove the affect from the discussion and to use figurative "as if" language. As anxiety increases in a group, so does the potential for metaphoric language. It becomes a safety valve for group expression. The metaphor is used by group members to communicate situational difficulties,[7] indicate group resistance,[8] confront group leaders,[9] confront group members,[10] reveal personal identities,[11] promote insight,[12] and provide future direction for the group.[13]

CHAOS

In the language of chaos, the group metaphor is the strange attractor of the group's collective unconscious. It reflects the pattern or topography of the group's psyche organization. The pattern is ever-changing, never repeating, constantly evolving as the group works to resolves its issues and unfold its future. At any one moment, the emerging metaphor provides a glimpse into that future; however, any interpretation provides only one of many possible meanings. As Jung realized, "The symbol is alive only so long as it is pregnant with meaning,"[14] and any attempt at defining it discharges its energy. Metaphors are constantly stretching and folding, appearing and disappearing, as the group randomly searches its collective psyche for resolution to immediate issues, while simultaneously organizing itself for the future. I imagine the collective mind weaving beautiful strange attractor patterns that are pictures of the metaphors. Group metaphors provide us glimpses, albeit small, of the collective mind organizing itself. Eventually metaphors, if properly contained, differentiate from the collective psyche and emerge at the manifest level of the group. Here again, the ability of the group leader to construct an effective container that enables the group to tolerate ambiguity and frustration permits metaphoric development to unfold fully, without the group becoming emotionally overwhelmed and thus settle for a regressive solution.

The leader is charged with the early group responsibility of identifying unifying themes from among the different narratives that individual members bring to the group. Group metaphors provide the clues to creating a successful group vision to carry the group through the initial stages. The vision becomes the attractor that provides meaning and pulls the group members toward it, building a sense of cohesiveness. During ascension of the arc, group metaphors provide generative meaning and creative opportunities for the group. Generative metaphors are discussed in chapter 10.

HOW TO RECOGNIZE A GROUP METAPHOR

There are three characteristics of group stories that indicate they are serving as group metaphors.[15] First, stories from outside the group are often distorted to conform to the present group situation. Second, the language used in these stories appears drawn from the current ongoing group. Third, the "characters and plot of the story frequently correspond with events and relationships in the immediate group."[16]

FUNCTIONS OF METAPHOR IN GROUPS

Comprehension of metaphoric language can provide therapists with an additional leadership tool as well as information about the group. Some methods of utilizing metaphors follow.

1. It enables the group leader to verify the stage of group development. It is well documented that a developing group moves through discrete stages with specific group behaviors defining each of these stages.[17] The leader has the responsibility of guiding the group through these various stages of development. At times this becomes very confusing, particularly if there is no coleader with whom to discuss the group's progress. Metaphoric language serves as a map for the therapist. Given the unstructured nature of most groups and their resultant ambiguity, a vehicle such as metaphoric expression is unconsciously utilized to determine a course of action. Metaphoric language becomes the staging area for preparation of the group's future course of action. It symbolically enables the group to try out new behaviors, ideas, and feelings, helping them to decide on norms for their future interactions. Careful attention to metaphoric expression helps the therapist navigate during these times of transition.

2. It provides information about members' identities. Metaphors provide members with an indirect and safe method of self-disclosing without incurring the personal responsibility that often accompanies such disclosures. "Members frequently encode important messages about themselves in statements they make about themselves or in stories they tell about animals or other characters; . . . the metaphoric framing enables one to make important statements about one's own identity or the identity of another while at the same time denying that a serious statement was made."[18]

3. It provides a method for directing the group's attention from a past- to a present-centered focus. The unique healing power found in groups stems from its potential to reflect the larger world; that is, the group represents a microcosm of each member's personal reality and thus provides a more realistic environment for counseling. As members participate in an ongoing group experience, it becomes evident that the issues and concerns that brought them to therapy are soon reflected in their group behavior. Successful group work integrates the members' cognitive understanding of their particular concerns with the experiential awareness of their reenactment of these very issues within the group. This integration addresses the two primary domains of group therapy: a past (there-and-then) focus, versus a present (here-and-now) focus. Both are necessary for successful resolution of client problems. Illumination of metaphor provides the

bridge by which group members move from a cognitive there-and-then understanding of their problems to a more here-and-now experiential awareness.

4. It offers a method of creatively generating feedback about the group's processes. Two ingredients have been suggested as essential for effective group functioning, here-and-now focus and process commentary, which were discussed earlier in the book. Illumination of metaphoric material generated by group members stimulates creative ways for the group to examine its ongoing interactions.

METHODS FOR UTILIZING A GROUP METAPHOR

Once a metaphor is recognized, the group leader can use it as information, amplify it, illuminate it, or interpret it. First, a metaphor provides information about the group's progress. The leader can treat this information as feedback and influence the direction of the group accordingly. Second, the leader may amplify the metaphor (e.g., continue within the framework of the story and offer resolutions for the group's dilemma). Third, he may illuminate the metaphor and ask group members to provide meaning for it. Finally, he may choose to interpret the metaphor and suggest possible meanings for the group's current situation.

The aim of metaphoric analysis is to utilize the information contained in the metaphor to facilitate the group's development. Therefore, it is not always necessary to call the group's attention to the metaphor. However, as the group develops teaching members, the benefits of metaphoric analysis provide them with yet another tool for therapeutic insight and personal growth.[19]

There are two simple guidelines to follow when working with a metaphor. In the beginning stages of group, the leader should proceed cautiously and treat the metaphor as feedback about the group's progress. Later, when high levels of cohesion exist, illumination and interpretation might be used because members are better able to tolerate the anxiety associated with directly struggling with the metaphor's manifest meaning. Each of these methods of utilizing group metaphors has validity, but the preferred intervention strategy is amplification of the metaphor. Amplification of the metaphor is less threatening to group members because it maintains the safety of symbolic language and avoids the reduction required by interpretation or illumination. It is impossible to know exactly what message is being conveyed by a metaphor.[20]

Metaphors serve vital functions throughout the life of a group. In addition to offering retreat from emotionally charged situations, they transcend the limits inherent in language and permit a group to rapidly process complex and diverse information, consolidate ideas, and generate solutions to group problems, as well as prepare the group for its future. Therefore, careful deliberation should be undertaken before deciding to interrupt what appears to be a natural part of a group's functioning. Three related areas should be assessed before deciding to illuminate or interpret a metaphor: group safety, clarity of interpretation, and leader intentionality.

First, related to group safety, metaphorical language occurs in response to group situations that are too risky or threatening to address openly. In these instances, the group moves to the latent level, attempting to distance itself from the affect while working to resolve the issues. The leader should not attempt illumination or interpretation of the metaphor until the group has made repeated efforts to resolve the issues at the latent level. If strong resistance follows such an intervention, further latent level exploration of the metaphor should be considered. Second, interpretation of a metaphor reduces it to language, which is inadequate to convey its full meaning. Unlike metaphor, "Language is often insufficient to capture the intricate structure and dynamics of the group-as-a-whole."[21] Any interpretation offered by the leader may impede the natural therapeutic process by calling attention to only one aspect of the metaphor. "No one ever completely understands all the ramifications of the metaphor as it is occurring—the images resonate and reverberate in the unconscious, and the work being done continues, even after the meeting is over."[22] Interpretation should be restricted to those occasions when the group has exhausted the metaphor's potential at the latent level.

Third, the leader should consider his intent in interpreting the metaphor. Answers to the following questions can provide insight into the motivation behind any interpretation and may help in deciding the appropriateness and timing of a metaphoric intervention: Who is best served by the interpretation? Will the interpretation have a negative impact on any of the members? Does the interpretation serve to facilitate the group's progress? Can the group be better served by allowing the metaphor to continue uninterrupted? The benefits of group counseling are negated by a leader who is too directive or persuasive. Members are inhibited from acting, sharing insights, and, in general, assuming responsibility for the group's functioning and ultimately their own.

Leaders must also exercise caution in interpreting metaphors. Forcing examination of the metaphor before the group is ready can have deleterious effects. It can inhibit the group's progress and their willingness to self-disclose, and it can alienate the leader from the group. Mishandled metaphors might even cause members to be absent from group sessions.[23] A metaphor is a privileged communication between a group member or the group and the leader and should be accorded all the rights of such an interaction. Only after all indirect means of exploration have been exhausted should illumination or interpretation be attempted.

The amount of time spent working with the metaphor depends on the level of anxiety present in the group. Remember, the metaphor acts as a safety device that offers group members retreat from highly emotional situations.[24] If the group responds to the metaphoric intervention with confusion and misunderstanding, wait until a safer group climate develops. In some cases, when the leader has attempted a direct intervention and the group resists, experimenting with amplification may allow the group to proceed with the metaphor. The following examples of group metaphors selected from various groups highlight the preceding discussion.

A New Group

During the first session of a new group, one member talked about the recent birth of his daughter. He shared his enthusiasm about the birthing process and the novelty of the experience, as well as his fears and concerns about infancy. Other group members added their own experiences of raising children. At the individual level, the member began to reveal personal information about himself; additionally, he was sharing something deeper and more immediately relevant, that he was both excited and frightened by the newness of this group experience. At the group level, within the symbolic safety provided by the metaphor, members, too, were able to express their trepidations about this new experience.

The facilitator has several choices at this point: (a) simply note the metaphor and allow the group to continue; (b) bring the group into the here-and-now by asking the member if he is both excited and frightened about what might happen during the course of this group; or (c) ask the group if there are any similarities between what is being discussed and what is happening in this group right now. Since the choice often involves a variety of factors, the facilitator must trust his or her own instincts as to what approach to take, keeping in mind the timing and level of group trust that exists.

The Faulty Muffler

Midway through the second session of a group, one member told of the difficulty she experienced when the muffler fell off her car earlier in the week. She expressed her frustration and embarrassment about the noise her car was making. During the first meeting, she had revealed a considerable amount of personal information about herself that had gone unchecked by the facilitator. The latent level of her communication addressed her concern that, the previous week, she had revealed too much about herself and was feeling frustrated and embarrassed.

Again the facilitator has several choices; he or she could (a) bring it to the here-and-now by asking her if she felt she had revealed too much about herself in the previous session; (b) ask the group to explore any similarities that exist between her story and the group; or (c), requiring more skill, develop the metaphor about the muffler. Exploring ways she might have prevented her muffler from falling off and/or discussing her feelings about that event might provide her a less threatening opportunity to resolve her anxiety.

Snow Storm

In the following example, the metaphor documents the group's stage of development. Before the fifth session, a group had been experiencing the facilitator as cold and distant, responding to them primarily in a cognitive mode. At the beginning of this session, one group member, seated to the immediate right of the facilitator, related the following story:

> I sometimes feel that if I were in a cold, winter, snow storm, a complete white-out, and my car was not working, I would have to find a way out of that situation by myself. I would either have to fix my car or find a way out of the storm. It would be very difficult for me to ask for help.

The facilitator that evening was wearing a white dress. At the manifest level, the individual was revealing something about himself, his difficulty in asking others for help. However, at the latent level, he was confronting the facilitator (*white out*), with her lack of affective expression (*cold*), and her inaccessibility to the group (*very difficult to ask for help*), and moving the group toward Stage 2, conflict and confrontation.

The facilitator could ask the member if he has been experiencing her as cold and emotionless or ask the members to explore possible meanings this story may have for this group. Metaphors confronting the facilitator are often difficult to comprehend because some blind spot in the therapist's personality is usually activated; however, time spent in reflection, after the therapy hour, often produces insight into its meaning. Remember, if the group feels stuck, examine the metaphors for clues to the "stuckness." Stuckness and lack of direction are often reflected by an increase in metaphoric expression.

> Fantasy and metaphor as well as other novel approaches in the use of language seem to be natural responses to . . . ambiguity. These fantasy episodes tend to occur when members fail to have a clear sense of what is going on in the group and when communication and understanding seem blocked.[25]

What to Wear

Another example illustrates how one metaphor can develop throughout the lifetime of a group. During the first session, a member introduced the theme of dress as a metaphor for working or not working in the group. She related an incident that occurred several weeks before this group meeting. Heading home from work, she had a flat tire that she was reluctant to change because she was dressed up and fearful of ruining her good clothes. That evening, in the group, she was wearing a dress.

The following week she wore an old sweatshirt and blue jeans. One member commented that she came dressed to work tonight. Throughout the lifetime of that group, dress served as a metaphor to indicate member willingness to work within the group. During the seventh session, the metaphor was used by the facilitators to assist a member who was struggling with ways to change her behavior at work. She was feeling overwhelmed and stuck by the amount of change she felt necessary to enhance her relationship to her job, particularly with her boss. The facilitator suggested to her that, instead of immediately buying a whole new wardrobe, she should consider trying on one piece of new clothing at a time.

The following example demonstrates how a metaphor can develop over the lifetime of a group.

The Group

The group was comprised of 10 women (8 participants and 2 coleaders) who were part of a self-analytic group designed to help students learn about group dynamics and interpersonal behavior by studying themselves.

The group was co-led by two women who were members of an advanced group class. They had both completed the beginning group course and had led two other similar groups. There was an absence of a formal group agenda, and the leaders' roles were to promote intermember communication and provide safety for members by monitoring the group's development while encouraging the members to provide their own direction.

Giving Birth. The group was very tentative during the first two sessions; they were searching for meaning and structure. Repeated attempts to have the leaders set the agenda were frustrating for group members because anxiety levels were high, and members were uncertain how to proceed. Susan, a member, began the third session by informing the group she was willing to take a risk and share something of herself. She told the group she had been giving serious thought to getting pregnant and having a child next year (she was neither married nor involved in a current relationship). She asked them for their reaction to her decision. The group responded politely, but it was evident that several members were upset by her decision, which they perceived to be irresponsible (confirmed by their weekly journals). The one issue that was not discussed by the group, surprisingly, was who would father this child.

The group members moved cautiously during the next several sessions as the leaders urged more involvement through sharing and self-disclosure. During the fifth session, another member commented that all the "pushing and pulling" by the leaders felt like the group was in "labor."

Susan began to push group boundaries by confronting members with their tentativeness about fully participating in the group. Conflict began to surface as group members cautiously asserted themselves and confronted the coleaders. A third member noted that the group members were beginning to take "baby steps" toward controlling their own futures. The conflict continued, and, during the next two sessions, group members openly confronted the leaders for their lack of leadership in setting the group's agenda. Throughout the tenth and last meeting, the group members were able to share many of the frustrations they felt during earlier sessions when the group lacked specific direction on how to proceed. Members shared that the group was finally beginning to work and were regretful that it was ending. At the very end of the session, one member turned to Susan and asked her how she felt about the group now. She responded *"I feel like a mom."*

Analysis. Susan's concern about getting pregnant served as a metaphor for this group, a vehicle through which members communicated their concerns about the group without exposing their accompanying vulnerable feelings. Susan, who had previous group experience, became the spokesperson for the metaphor, and other members contributed to its development as the group progressed. A literal translation of the metaphor reveals that the members were questioning how they would form as a group (become pregnant) without direction and guidance (father figure, i.e., structure).

In addition, they were confronting the female cofacilitators and asking if they had the necessary skills (potency) to help the group develop (impregnate the group) into a cohesive, working unit (give birth). The question asked of the leaders was, "Will our group develop without a strong, guiding leader (the father figure)?"

The metaphor expanded (pun intended) and tracked the group as members became less anxious and more secure in their group roles. Although the group was still struggling at the half-way point, the reference to labor suggested that they had indeed found some direction (become pregnant). However, it was still too unsafe to openly and directly address the leaders with their concerns. Later, as the group became more predictable and as uncertainty diminished, conflict surfaced and was expressed.

The member characterization of the group as taking "baby steps" confirms that this emotionally charged issue was beginning to be addressed at the manifest level.

Finally, group members expressed their frustration and confronted the leaders over the lack of direction provided for the group. Confirmation that the group had shifted from the latent to the manifest level occurred in Susan's closing statement, "I feel like a mom."

When groups struggle and avoid addressing emotionally charged issues, examination of group metaphors can provide clues to the "stuckness." In many instances, the metaphor offers suggestions for resolving the dilemma. Clearly, this group was experiencing some anxiety that was communicated by both the use of metaphor and its message. In this case, the leaders could have used the information from the metaphor to provide more direction for the group. Although members eventually took control, some of their anxiety could have been abated by careful attention to the metaphoric language. The group would have blocked interpretation or illumination of the metaphor. In the third session, one of the leaders asked the group members if they thought this discussion of pregnancy and having a baby had any meaning for the group's present situation. The question was met with silence.

The leaders could have chosen to explore the group's anxiety by amplifying the metaphor. One possibility may have been to discuss the issue of mother-headed, single-parent families. Subsequent discussion could have addressed particular problems faced by single-parent families when mothers are confronted with dual-parenting responsibilities. Other discussion could have focused on contrasting styles of parenting: one that encourages children to become self-supportive and responsible for their own development, another that is more directive, and a third that is some combination of the two. Amplification of the metaphor may have served as a nonthreatening method of exploring the positive and negative consequences of various leadership styles.

Almost without exception, illumination or interpretation of the metaphor during the early stage of a group's development is met by the members with resistance in the form of confusion, withdrawal, and puzzlement. Even in the later stages of a group, after high levels of trust have been established, there is still reluctance to explore direct interpretations of individual or group metaphors.

PARENTING

In the first two sessions of a training group, the coleaders allayed their own anxiety by dictating rules to the group about being on time, using "I" statements, subgrouping, operating in the here-and-now, and maintaining confidentially. The imposition of these rules, without the benefit of discussion and process by group members, stifled the atmosphere. In supervision, the group leaders were given feedback about their authoritarian behavior and its resultant effect on the group. During the third session, they sought to redress their behavior by soliciting feedback from the group members on the imposed rules and the manner in which they were presented. Group members were very hesitant to comment directly on the leaders' behavior and offered comments such as, "I hadn't felt one way or another about the rules. It seemed fine to me." Several members responded that they could not remember what the rules were. When pressed on this point, the members were surprisingly able to recall the rules in full detail. No comments were offered on the manner in which the rules were imposed on the group with the exception of one member who said she felt "It was necessary for the leader to take charge." Shortly after this exchange, a group member related the following story. It pertained to a high school student who had been on an athletic team he coached. He told the group how little support the student had gotten from his mother. She felt athletics were a waste of time and wanted her son to be more productive and find after-school employment. She made it difficult for him to practice by prescribing a list of rules he must follow in order to continue with the team. The coach felt she was totally unsupportive of her son, and he (the coach) was unable to approach her effectively because of her dictatorial manner.

Other members began sharing similar experiences about nonsupportive parents, particularly the traditional role of mothers, which is usually one of nurturance and support. It should be noted here that this particular group was being co-led by two women. One of the leaders illuminated the metaphor by asking the group if they felt there was any relationship between the story and this group experience. The silence was deafening. She then attempted an interpretation of the metaphor by exploring her role as the perceived mother of this group. This was met with confusion on the part of two members, and the group quickly returned to the safety of the metaphor.

A less threatening method of utilizing the metaphor would have been to stay within the narration and amplify it, perhaps exploring the difficulty mothers have in providing structure for their children while at the same time creating a supportive environment for them. In addition, the leader could have used the safety of the metaphor to explore how group members felt about mothers (referring to the group leaders) who were perceived as controlling authority figures. In fact, at one point, while participating in the ensuing discussion, the leader offhandedly (not as an intended intervention) stated that she felt many mothers had demands on them that children could not appreciate. Shortly thereafter, one member said to the coach, "You know, maybe that mother was doing the best she could." There is considerable agreement that working within the therapeutic framework of the metaphor is more effective and productive than attempts at interpretation.[26] "Discussion and intervention within the metaphor takes place on different levels of reality at the same time, embodying both the concrete patterning of events and their symbolic equivalents."[27]

ON TASK

The following example demonstrates a method for amplifying the metaphor and underscores the effectiveness of such an approach. The transcript that follows was excised from the 11th session of a personal growth group of high school peer counselors who were being videotaped. They had been carefully informed of the group requirements, and each had volunteered for the experience. The group was co-led by a school counselor with considerable group experience (Bob) and a graduate student (Julia). The group was confronting the leader's (Bob's) preoccupation with the task: the technical aspects of the videotaping, the seating, camera placement, microphones, and his lack of expressed concern for the group members. Therefore, the group members began to explore their own obligation to the group, which required about 4 hours of their time each Monday night, including travel to and from the television studio and dinner. The group began the session by replacing the straight-backed chairs used in previous weeks with sofas and chairs that were stored in the television studio.

Cathy: It feels more comfortable with these chairs.

Tim: It feels like . . . someone's living room.

The group members began to talk about their struggle with wanting to spend more time with their friends although feeling obligated to be with their families (i.e., obligation to the group).

Tim: It is difficult deciding between them.

Bob: (*staying within the metaphor*) Family is more of an obligation, more of a duty.

Tim: When something has to be crossed off my list, it is usually family.

Molly: It is scary.

Bob: Do you feel your parents (using the word *parents* to represent authority figures, e.g., group leaders) have an appreciation

	for the struggle you are going through? What do you do with your parents?
Tim:	I ignore them a lot. Just lately I'm aware of it. When I was younger, I was afraid to go out, to abandon them. I don't talk to them much anymore. I don't look at things from their point of view, they don't know what's going on with me.
Bob:	What about your dad? (*Within the metaphor, the leader begins to explore how Tim views his role in the group.*)
Tim:	In a way I feel real close to my dad. But at the same time he's a guy I don't think in a lot of ways I'm very different from and some ways we are alike. There are some things I can talk to him about, and others I don't think he'd understand . . .
Julia:	(*coleader*) Have you thought about trying once with your dad? Giving him a chance?
Tim:	I've learned through experience. I have already tried, but maybe it's time to try again.
Bob:	It sounds to me like you're trying. This struggle, leaving home, moving away. Also I hear you saying that you have ignored him . . . How do the rest of you feel? (*looking at Julia*) Do you feel that same struggle with your family? (*now working to include the rest of the group in the metaphoric discussion*)
Julia:	I guess our family doesn't always get together. My sister comes home on Sunday and I try to be there to spend time with the whole family. It's hard to balance between family and friends.
Bob:	How do you handle your mom and dad (e.g., leaders)? I hear how Tim does it. On one hand you kind of do what they want on the other hand you are beginning to do it your way.
Julia:	I basically agree with what they believe so there's not much conflict. I do wish we were a lot closer. It's hard to talk to them.
Bob:	It's certainly hard if you'd like to get closer, to say you'd like to get closer.
Julia:	Yeah, especially with my dad.
Cynthia:	My dad is too, kind of weird, in a way. My dad and I are alike. . . . He likes to be around people and so do I. . . . But he just changed jobs, like he went from doing something, he's a lawyer and he didn't want to go back to private practice, but he didn't know what else to do, so he's really stressed out. Last night he asked me four times how school was. I'm like,"Dad you already asked me," and I felt real bad. I wonder if he just doesn't remember or if that's just the only thing he can think to ask me. I feel kind of bad if it is. . . .

Cathy: (*who has gotten a lot of support from this group*) My family is up there. My family has been really important to me . . . they're always there for me. I find myself sometimes crossing them off too and I feel kind of bad.

Bob: It's like somehow you know your family is always going to be there so sometimes its okay that you take them for granted. But maybe not for much longer. You're at a point in your lives where that relationship with the family ends, at least the way it is now. (*Group was ending in 2 sessions.*)

Cathy: Those rebellious teenage years.
(*More discussion*)

Bob: (*to Cynthia seated to his left on the sofa*) I was thinking while you were talking earlier, if you could say something to your dad or ask him something what is it you'd like to say to him? (*within the metaphor, asking her for direct feedback on how she perceives his role in the group*)

Cynthia: To my dad?

Bob: You were saying you felt that . . .

Cynthia: I just want him to be happy, not be so stressed out. I'm afraid he's going to have a heart attack. I just want him to be more relaxed.

Bob: So you don't feel like you can pursue that with him.

Cynthia: I really like when my sister Ann comes home and he does too . . . Because my dad and my sister have a close relationship. One thing my dad is trying to let me know right now is just that it's okay that I'm not doing the things my sister, Ann, did.

Bob: Somehow he is letting you know . . . that it's okay to be different from Ann.

Cynthia: He is, he's really sweet, he really is . . . Like my birthday he left a note on the refrigerator that said "Happy Birthday, Molly" and went on to say that all great people were born in September. His birthday is in September, too. He lets me know he cares. It's just when he's stressed that I don't know what to do.

Tim: I can identify with Cynthia so much when she's talking. . . . My father is an attorney, too. So we're kind of in the same situation. I think about how hard my dad works. He works harder than anyone should work. And I know he works for me, so our family has a nice home, things like that. . . . He works hard and I really want to let him know I'm grateful and it's hard for me. I think he knows. I think we kind of share that, we don't talk a lot. I know it's hard for my dad to talk to me, too.

Bob: (*in response to Cynthia*) It's funny how dads write letters and write things down . . . somehow you don't feel they can say it. (*The leader had been writing and sending weekly summaries of the group to the members. The letters had a more caring tone than the group was obviously experiencing from him in person. Indirectly, through the letters, he was letting the group know he cared about them.*)

Cathy: That's kind of like my dad. I mean we never talk. But in ninth grade I was having a lot of trouble. . . . my dad knew it and it was kind of weird he bought me this stuffed animal. He just gave it to me and said that he hoped everything would work out and I was just overwhelmed 'cause he just never showed much affection or anything. Ever since then we've gotten along a lot better. Sounds like a fairy tale doesn't it? (*Here again is the reference to an indirect means of caring, i.e., He does things for me so he must care.*)

Bob: My own father and I never had a good relationship. I could never live up to his expectations. He was a driven man. And he had a stroke at 67, 5 years ago. He has changed in the last few years and it took me 4 years to realize. . . . He's become a very loving man with a real spiritual quality that I never saw before and it was interesting that he had been changed, but only in the last 6 months have I seen him differently. It was me. There he was wanting to love me now and I was still holding onto that picture I had of him a long time ago. I really found out that it was me who was holding him out of my heart. A real shock to me . . . I have tried to understand his struggle. To see his side that I never saw as I was growing up as his son, being so absorbed with my world. (*Within the metaphor, the leader self-discloses about his own struggle as a son.*)

(*A long silence during which members sat quietly and drew inward, avoiding eye contact with others in the group*)

Bob: But it was real hard for me to tell him how I really felt, 'cause he was always in charge. It always seemed to me that it should come from him to me. And it was really me who had to see him differently. (*continuing within the metaphor, reiterating how difficult it is to directly confront father, i.e., group leader*)

Julia: I'm just wondering if you are not asking people here to see you differently than they see you. (*interprets the metaphor and the group responds by discussing it at the manifest level*)

Cathy: (*to Bob*) It's funny you say that because I do see you differently. You seem a little more interested in us. . . . I mean more in the person than in how the group is progressing the way it should.

Bob: I think I have seen it tonight and it feels different to me, a real sudden awareness.

Julia: Yeah, I've seen a side . . . here tonight. I like it a lot better than how you were.

Bob: Thanks.

Jan: It's weird discussing this tonight how it has changed. It's neat.

Julia: Are you feeling different?

Jan: Yeah, I really like it a lot. I'm wondering, like, beside the topic of love, the chairs feel more comfortable, more at ease. . . .

Bob: It's like being in a living room. (*reference to family i.e., group*)

Jan: At home.

Cathy: I haven't noticed the cameras once tonight.

Cynthia: I really like it. I just feel so comfortable sitting here. It's not like a group. It just seems like I'm . . . talking to a bunch of my friends . . . at my house.

Tim: (*sitting between two group members*) I feel so enclosed by this sofa. The other chairs always felt like you could just fall off.

Jan: It's like you really want group to continue. (*to Julia*) Can you talk about how you get along with your dad or is that too personal? (*Returning to the latent level, Jan asks the second facilitator about her relationship with the male leader.*)

Julia: No, I think we've really struggled throughout the years there have been times when we have been close. I don't know that I've ever felt filled by my dad. I know he has loved me dearly but he doesn't always know how to say it. I just want more than I get.

Bob: (*to Julia*) Dads have such a powerful influence over us. (*a reference to her role as cofacilitator and graduate student*)

Julia: (*to Bob*) Sometimes you remind me of my dad. Sometimes I feel close and sometimes distance. (*manifest level*)

Bob: (*responding at the latent level to the whole group*) It's hard being a father so responsible. You have to take care of the family and take care of everybody and make sure that everyone is okay. The responsibility gets in the way and care and love get expressed through it rather than directly. (*turning to Cynthia*) That's why it's nice sometimes that kids can figure it out, what he really means.

Tim: I just feel so bad. It just took me till last weekend to figure that out. I just feel like I wasted so much time.

Bob: It took me 40 years to figure it out with my dad.

Julia: Maybe we want to hold onto what we have left because it (*group*) will be ending for us soon.

Cathy: . . . it should go on. It seems like it just started.

Bob: A student wrote in a paper for me once "just when you figure life out it's over."

Amplification of the narration and staying within the safety of the figurative language allowed for multiple processing. In the preceding example, the group members were able to simultaneously address family of origin and group issues. The group was effectively able to confront the leader, thus seeing another side of him, all within the safety of the metaphor.

This group demonstrates how it's possible, through the use of metaphors, to undergo significant change while operating at the edge of chaos or complexity. In this example, the coleaders effectively work at the group's latent level, maintaining the fine balance necessary to push the group forward while preventing extreme turbulence. Working with the collective mind is an art that depends on a finely tuned ear to hear the group's collective voice that is expressed through metaphor. The metaphor is not static and, as the group develops during this session, so does the metaphor. Tracking this dual evolution of the group and the collective metaphor requires an ability for parallel processing by the leader. This skill is very necessary in the later stages of group development.

This chapter introduced a power tool, the group metaphor. It also identified some areas of concern associated with the illumination and interpretation of spontaneous metaphor in groups. Caution was urged in directly interpreting metaphors. The preferred method introduced was amplification as a more natural and less intrusive means of metaphor in groups. It is hoped that, as group leaders become familiar with the therapeutic potential of metaphors, they remain sensitive to its function, power, and complexity.

Author's Note: Some of the material in this chapter previously appeared in B. A. McClure, *Counseling and Values* (Vol. 38, pp. 76–88). Copyright © 1994 by the American Counseling Association. Reprinted by permission of the American Counseling Association.

Regressive or Limit Cycle Groups

A group is impulsive, changeable and irritable. It is led almost exclusively by the unconscious. . . . Though it may desire things passionately, yet this is never so for long, for it is incapable of perseverance. . . . It goes directly to extremes; if suspicion is expressed, it is instantly changed into an incontrovertible certainty; a trace of antipathy is turned into furious hatred.[1]

This description of mob behavior characterizes the regressive and psychologically immature aspects of groups. Groups are primitive, impulsive, unintelligent, restrictive, and potentially destructive. In groups, members remain unconscious, minority voices are repressed, and internal conflict stays unresolved.[2] The psychological forces that keep group members stuck in the early stages of development have been described as ". . . threats to the loss of one's individuality and autonomy, the revival of early familial conflicts, and the prevalence of envy, rivalry, and competition."[3]

This chapter examines the dark, denied, and unacknowledged behavior of groups and organizations. Of particular interest are those groups where members remain dependent on the leader for direction, ignore or deny their negative attributes, and lack a common positive identity or sense of esprit de corps to bind them together. These groups are labeled regressive.[4] Typically, regressive groups are stuck in the forming stages of development and exhibit four general characteristics: (a) avoidance of conflict and dissent, (b) abdication of responsibility for the group's behavior and dependence on the leader, (c) group narcissism, and (d) psychic numbing. When combined, these traits inhibit the maturation of groups and result in a kind of "group mindlessness" wherein members distort their inner and

165

outer realities to conform with a dominant group view.[5] This chapter examines the development of regressive group characteristics and offers suggestions for transforming these rigid groups into more productive organizations. Finally, leader behaviors that can liberate regressive groups are considered.

CHAOS

From the perspective of chaos theory, these groups are stuck in limit cycles. They do the same things over and over again. The group must, of course, endure chaos if it is to develop. However, regressive groups have not developed sufficient fixed points to which they might anchor to contain themselves as they experience the anxiety and upheaval necessary for reorganization to the next stage of development. The group shuts down, limiting energy exchange or feedback from other systems. As the group becomes more isolated, opportunities for constructive reorganization decrease. The limit cycle, or back-and-forth motion of the pendulum, characterizes the group's movement as novelty is damped. If self-organization is blocked, the group still changes; however, the resulting structure is not self-generating or self-maintaining.[6] Group members, similar to the slowing pendulum, move toward the middle or collective mean as energy dissipates and difference is diminished. Thus, the collective or spiritual potential present in the preforming stage is thwarted.

In order to illuminate the discussion and highlight the manifestation of regressive group characteristics, two examples are provided. A consulting assignment provided me with the first example, a college faculty department. This group clearly illustrated the predominant form of regressive groups. The second example is an examination of the behaviors of our collective group, the United States, during the Persian Gulf War. Regressive group characteristics are most pronounced when threats to a group's survival, actual or perceived, exist.[7] This latter example best illustrates how shadow material, during times of increased group stress, is activated and manipulated to serve the needs of the leader.

All groups contain a collective shadow consisting of the unexpressed emotional negativity that group members experience as threatening. Additionally, personality characteristics and emotions that members are unable to accept in themselves are also hidden in the collective shadow.[8]

As noted earlier, groups develop through a series of stages. Beginning with the Disunity stage, effective groups resolve conflict, develop a sense of "we-ness" or belongingness, become highly cohesive and productive, and, for some, end in a termination stage. However, many groups never sufficiently work through their differences and evolve beyond the initial stages of development. Not only are these groups unable to maximize their full creative potential, but they develop destructive or regressive tendencies that remain largely outside the awareness of the group's members. While most groups exhibit regressive tendencies at one time or another, healthy and mature groups are able to endure periods of high anxiety without regressing to destructive or injurious behavior to themselves or others. What distinguishes the mature group from its regressive counterpart is its ability to openly express conflict, encourage divergent thinking, and nurture diversity and difference among its members.[9]

As groups or organizations develop safety and trust, a collective identity emerges.[10] However, the level of intimacy established among members in these groups depends on at least two factors: group size and purpose.

Counseling or therapy groups foster high levels of self-disclosure and intimacy among their members. Group size is usually limited to 6 to 12 members, which facilitates the intimacy-building process.[11] Task-oriented groups or organizations range in size from several members to several hundred members. These groups usually meet on a regular basis with a specific agenda that focuses almost exclusively on conducting the business of the organization. Very little time or effort is directed at maintenance or affective activities within the group. Large collectivities such as neighborhoods, communities, or nations may share common characteristics based on ethnic or religious or geographical identification but never hold a formal group meeting. Although there is considerable debate about the nature of large groups, there is common agreement that they manifest a collective identity and at times share a collective agenda.[12]

The larger the group, the more difficult it is to address and resolve member differences. In fact, many argue that large groups or crowds are inherently regressive.[13] Hence, the individuals who comprise them are prone to relinquish their individual identities in favor of feelings of anonymity, unaccountability, and invincibility, thus making them more susceptible to "collective behavior."[14] But in large groups, for example, it is often the group leader who can promote tolerance, cultivate well-being, and invite the expression of conflict and dissent. In a sense, the actions of the group leader can greatly influence the expression or repression of shadow material.

This chapter examines the regressive characteristics of groups, large or small, that hinder the group's creative and productive potential. Any group leader can benefit by understanding the factors that contribute to the development of regressive groups.

Groups move through various stages of development, with each stage serving as a platform on which the next one is erected. Hence, it is critical that the early stages or foundation are soundly constructed. Keep in mind that initial group safety is developed by the boundaries and rules of behavior that are established by the group leader. As the group progresses, these rules are either adopted by the group or renegotiated with the leader. During this negotiation process, increased levels of trust and safety are engendered. Once a modicum of comfort is established, groups move into a Conflict/Confrontation stage.

THE CONFLICT/CONFRONTATION STAGE

The expression and resolution of conflict is stressed in this book as the most critical period in group development for several reasons: The group must untangle its preconceptions of the leader, resolve dependency issues, develop a norm for successful conflict resolution, and establish genuine cohesiveness. The group must come to terms with the leader. Members bring to groups preconceived notions of how leaders should act. Many of these perceptions are based on earlier experiences with previous authority figures, including parents. The most pervasive and unrealistic fantasy, albeit unconscious, is that the leader is omnipotent and will somehow parent, protect, and satisfy each member's need. Obviously, no matter how competent the leader, he or she fails to meet these expectations; thus members become frustrated and angry. The leader's inability to be real or vulnerable, admit mistakes, share the power by promoting autonomy and independence, and encourage conflict and dissent results in validating the omnipotent parent illusion.

Underlying this disappointment is a second critical factor that must be resolved in the conflict stage: dependency versus counterdependency, or members' need to be taken care of versus their desire for independence. Fostering the "omnipotent leader" illusion inhibits group members from reclaiming the power they have invested in the authority figure. The more willing members are to recognize the leader's limitations, the more able they are to realize their own potential and capabilities. Successful resolution of this dilemma initiates the shift of responsibility for the group and its functioning from the leader to the group members.

Group cohesiveness is primarily developed during this conflict stage. Together, as group members undergo this period of anger, frustration, and chaotic activity, the relationships among them are strengthened. This stage serves as the forge in which the group bond is heated to white-hot temperatures.[15] Successful completion solidifies cohesiveness, whereas ineffective or incomplete resolution results in a brittle and fragile bond.

Groups that fail to effectively address or resolve conflict stagnate. Their creative potential is severely limited. They remain stuck, are unable to handle conflict, and develop a kind of malignant cohesiveness that enables them to maintain the appearance of harmony. In many cases, these groups are able to function effectively, but they lack flexibility, and under stress their noxious behavior is exacerbated. Once a regressive group has remained chronically stuck for a period of time, their destructive tendencies become reified. Then, any perceived or real threat to their illusion of harmony (survival) is met with angry resistance.[16]

REGRESSIVE GROUP CHARACTERISTICS

There are four dominant traits that characterize regressive groups: psychic numbing, abdication of responsibility for the group's behavior and dependence on the leader, group narcissism, and avoidance of conflict and dissent.[17] H. Scott Peck, in his treatise on evil, called attention to these regressive group characteristics. Recently, we again witnessed this brutality in the former Yugoslavia as Serbian and Croatian soldiers committed acts of barbarism. To some extent, all groups contain one or more of these regressive characteristics but mature groups have acquired the necessary skill to bring to light shadow material that may be impeding its growth.

Psychic Numbing

Over time, regressive group members anesthetize themselves to contradictions in the group. This psychic numbing process deadens feelings aroused by agonizing decisions that group members must make that often contradict their individually held ethical and moral principles. When numbing is complete, member values become synonymous with those expressed by the group, enabling members to participate with little or no noticeable discomfort in the group's activities. However, the physical, emotional, and spiritual damage done to members is often considerable.[18] As group members repress and internalize the contradiction between their own ethical values and those of the group, illness, depression, and anxiety are often the result.[19] However, this numbing process that members undergo is critical for the continuation of the group. Members who are unable to conform or deaden their awareness are customarily shunned and ultimately excluded from the group.

Abdication of Responsibility for the Group and Dependence on the Leader

Abdication of individual responsibility in groups is a particularly dangerous trait. Members avoid leadership roles in regressive groups, thus abdicating direct responsibility for the group's actions. The inability to challenge the leader collectively during the forming stages inhibits the group from effectively redistributing the power. H. Scott Peck speculates that the role of follower is much easier than the role of leader. As such, he summarizes, one need make no decisions, plan ahead, initiate action, or risk unpopularity.[20]

The turning point in many therapy groups rests on the leader's ability to confront group members on their dependencies within the group. The leader must then relinquish the leadership position so group members can experiment with assuming that role and learn "how to exercise mature power in a group setting. . . . The ideal mature group is composed entirely of leaders."[21]

The group's reluctance to take charge is one factor, but more likely the leader's ignorance of group dynamics or unwillingness to relinquish control is what causes members' inertia. Clearly, the two are interrelated. The leader's need for control and group members' willingness to cede their own power frames the notion of dependence on authority figures.[22]

In many cases regressive groups are led by authoritarian leaders who, as Arthur Deikman describes, emphasize obedience, loyalty, and the suppression of criticism.[23] These leaders are often charismatic and are adept at manipulating their followers' idealism to serve their own ends. The members readily invest their power in such a leader because it fosters their dependency fantasy—the wish for the idealized parent—that was not abandoned during the initial stages of group development.

This fantasy, which remains beyond awareness in adulthood, is seductive because it cultivates the mistaken belief that one is protected and cared for. The fantasy is insidious and is a primary factor in group members' willingness to cede responsibility for the group's behavior. The leader becomes trapped by the illusion, too, because leaders must continue to fulfill the group's expectation of the powerful parent or the group may annihilate them. In fact, the leader also wishes to believe in the powers of an idealized and omnipotent parent.[24]

If members do not gain autonomy, participate in decision making, and assume responsibility for the group, they will remain in the role of the child with only one responsibility: obedience.[25] The dilemma, of course, is that, in the role of the child, members surrender their own critical thinking capacities, projecting them onto the role of the leader whom they follow blindly.[26]

Group Narcissism

Potentially the most destructive consequence of regressive groups is group narcissism. Groups are bound together by a force called cohesiveness. In productive and healthy groups, cohesiveness is generated from within as

a manifestation of group pride. Pride develops from the synergistic interactions, accomplishments, and successes that result when members value,
respect, and trust one another. The malignant form of group pride is
group narcissism.

Narcissistic groups develop cohesiveness by encouraging hatred of an
out-group or by creating an enemy.[27] As a result, regressive group members
are able to overlook their own deficiencies by focusing on the deficiencies
of the out-group. Ironically, it is often the group's own shortcomings,
frustrations, anger, rage, and hostility that are projected onto the out-
group.[28] Thus, characterizations of the out-group are more often reflective
of the disowned negative attributes of the in-group.

By splitting off and projecting outward their dark, shadowy side, regressive groups maintain an illusion of harmony. A public myth is created by
the group that disguises any internal conflict.[29] Members of regressive
groups often describe themselves in glowing terms. As members numb
themselves to the apparent contradictions between the public myth and
the group's true nature, their ability to perceive reality is severely diminished. The greater the pressure in regressive groups to suppress critical
thinking and deny their own dark side, the more likelihood dehumanizing
actions against their perceived enemies will occur.[30] Extreme examples are
the Nazis and Ku Klux Klan.

Avoidance of Conflict and Dissent

In order to protect their myth, regressive groups not only commit physical
and psychological violence against out-groups, but silence members who
seek to expose the group's own shortcomings.[31] Since regressive groups
are unable to successfully negotiate effective norms for the expression of
conflict and its resolution, they avoid it. Criticism is muted and disagreements are damped; access to information that challenges group beliefs or
its public myth often is severely limited. Without dissent as a corrective
and stabilizing factor, regressive groups, in the extreme, are capable of
vile acts of cruelty, coercion, and domination.

Because minority views are repressed and labeled as corrupt, the range
of opinions to which regressive group members have access are restricted.[32]
Oftentimes, group dogma is simplified into slogans that are mindlessly
recited, further curtailing critical thought.[33]

Nongroup views are discounted, and access to nongroup members is
limited. Eventually, regressive group members self-monitor their contact
with outsiders and self-censor reading or other materials that conflict with
the group's dominant view. This self-censorship lessens emotional discomfort, facilitates psychic numbing, and permits group members to tolerate
the coercion and punishment used to stifle dissent.

STRESS

Regressive groups maintain the appearance of normalcy under nonstressful conditions. However, there is little flexibility in these systems, and stress increases their destructive potential. In a situation of prolonged discomfort, we humans naturally, almost inevitably, tend to regress. Our psychological growth reverses itself; our maturity is forsaken. Quite rapidly, we become more childish, more primitive. Discomfort is stress. What I am describing is a natural tendency of the human organism to regress in response to chronic stress.[34]

Chronic stress is also caused by perceived threats to an organism's survival. When such threats occur, the group resorts to primitive, nonrational methods of defending itself: scapegoating, projection, and even violence. William Golding's classic novel, *Lord of the Flies*, provides an excellent example of the development of a regressive group. His story, which some suggest is a commentary on society, centers on a group of boys marooned on a tropical island and traces their regression to primitive levels of behavior.[35]

Golding demonstrates that under periods of extreme stress groups are capable of violent and barbaric behavior, ultimately leading to their own destruction. Many regressive group characteristics are reflected in this story: scapegoating, group narcissism, suppression of divergent thinking, inability to articulate feelings, inability to resolve internal group conflict, and self-interest. The story illustrates that ultimately two factors are instrumental in the perpetuation of regressive groups: ignorance and fear. Stress exacerbates dysfunctional behavior. Examining behaviors of groups under stress often discloses their regressive nature. Regressive group examples abound in academia,[36] business organizations, religious communities, governments, and larger societal systems.[37]

REGRESSIVE GROUPS: A CASE STUDY

The most prevalent type of regressive group embodies the following attributes: exclusively task oriented, stuck between the Disunity and Conflict/Confrontation stages, poor levels of performance by its members, and a strong sense of group narcissism. Nevertheless, under nonstressful conditions these groups function at moderate levels of effectiveness. In times of stress, however, the group's destructive potential is exacerbated. The following example clearly illustrates the predominant form of regressive groups.

The group was task oriented. Its primary function was to negotiate the month-to-month activities of the department. It never addressed the maintenance or emotional aspects of the tasks faced by its members. Over a period of years, the tensions, conflicts, and anxieties that went unexpressed caused what little group cohesion that existed to become malignant.

The group had no norms of behavior for addressing the inevitable differences and negativity that existed among its members. Subgrouping and scapegoating occurred as methods of dissipating the anxiety experienced by members. The pervasive atmosphere of the group was one of mistrust. As a result, the group never progressed beyond the second stage of development. The absence of critical and divergent thinking resulted in the deterioration of each member's level of professional performance. It steadily declined over the years until the mean level of performance was poor. Deviation from that norm wasn't tolerated.

The department existed in a rather closed and geographically isolated university. It served as the repository for many of the unwanted projections of other departments and thus was very functional for the university at large. When new faculty members were added, they were often shocked by the "harmony illusion" that existed. It was evident that there was little basis for harmony among this group. New faculty members were subtly warned to hide their competencies, particularly any research or writing they were doing.

The department was insulated from exposure to its new members by the retention process. New members were voted on yearly by tenured faculty members. The voting was presumably based on the candidate's professional development, his or her teaching ability, and service to the university and community. However, no objective procedure existed for evaluating these criteria. In fact, tenured faculty never talked with new members about their professional development.

A collusion of silence developed where the message was implicitly communicated to new faculty that they were free to do little or nothing if they chose, and would be retained as long as they did not become too competent or comment on the group's dysfunctional behaviors. Deviation from the group competency norm or exposure of the dysfunctional nature of the group, a form of professional suicide, would have been met with a non-retention vote.

New members subgrouped and discussed the contradictions that were apparent in the group as means of lessening their anxiety. Those who stayed and obtained tenure after 7 years had been so acculturated to the group norms that they were unwilling to confront the group's destructive nature. In many ways, they had become psychically numb in order to survive the psychological trauma from the many contradictions in the group.

Over the years, a group shadow developed from all the disowned and negative aspects of the department. By projecting the shadow onto other departments and faculty members in the university, a sense of cohesion was maintained.

The department was responsible for training teachers, but never examined or discussed their own teaching techniques. Members of the university community and many of the college's own students were appalled at the ineptitude that existed in many of the department's classrooms. However, many of the department's meetings consisted of faculty venting their anger and frustration at the poor quality of teaching that existed throughout other colleges in the university. Any negative feedback by students was discounted. Peer review was nonexistent. Accreditation reports were ignored or discounted. The department had effectively cut off any source of feedback.

In times of extreme stress, the group was able to insulate itself from criticism and attack the potential threat. A new dean of the college was hired and quickly became aware of the reputation of the department, which now included the former dean. During his first year, the new dean instituted open college forums, opportunities for faculty to meet and discuss the future direction of the college. Emphasis was placed on research and teaching. The department members were discursive and disruptive during the meetings. Afterward, they were openly critical of the dean. He became a new target of projection during their faculty meetings. During his second year, the open meetings stopped.

The department chair retired, and a recently tenured faculty member was elected to replace him. The new chair was clearly aware of the dysfunctional nature of the department, but during his early years remained silent. He was perceived by the older faculty as their heir apparent.

Shortly after assuming the job, he began to institute new policies. The department meetings were conducted according to Robert's Rules of Order. Objective criteria were established for retaining new faculty, and an explicit review process was put in place.

The former chair had never addressed student complaints about faculty incompetence. But the new chair responded to those complaints by informing faculty and placing letters in their permanent files. He found incompetence beyond anything he had imagined. The emotional pressure on him to conform to established norms was intense. Finally, when he realized that some very difficult decisions had to be made regarding the department's professional comportment, he resigned. He summed up the regressive and destructive emotional atmosphere of the department when he described his fear of what might happen if he pushed for change: "I felt like they would murder me."

This example illustrates many of the common characteristics of regressive groups: avoidance of internal conflict, creation of a group shadow, inability to tolerate self-examination or self-criticism, and abdication of individual responsibility for the group's actions. The members of this group, for their own psychological reasons, chose to live less than authentic existences. For the most part, they were good people who had become numb to their individual and collective potential. The shield of tenure and geographic isolation protected this group from exposure, but variations of this theme are carried out daily in many business and community organizations.

THE COLLECTIVE SHADOW IN WAR

A more universal and illuminating example would be to briefly examine the behavior of a group in which we all share a collective identity, the United States, during a period of national stress, the Persian Gulf War.

In 1990, the United States underwent a period of chronic stress caused by perceived threats to the country's economic survival induced by Iraq's invasion of Kuwait. Our response to that threat embodied many regressive group characteristics—group narcissism and scapegoating, suppression of divergent thinking and dissent, individual abdication of responsibility for the group's behavior, dependence on the leader—and unleashed violent and barbaric behavior.

Prior to the war, the country had endured a long decline in national pride. Malaise, disillusionment, and discontent were widespread. Reports of our economic decline, failures of our educational system, political scandals, and pictures of our decaying infrastructure were pervasive in the media.[38]

The country had recently been transformed from the largest creditor nation in the world to the largest debtor nation. Costs from the saving and loan crisis were multiplying daily. Homelessness, AIDS, crime, environmental decay, the national debt, trade deficits were juxtaposed against cries of freedom emanating from Eastern Europe and stories praising the advancing economies and educational systems of Japan and West Germany. For us, there was no end to the bad news. Our collective fantasies of national superiority were eroding. For many years, we had colluded with President Reagan to maintain his illusory parental image by allowing him to foster an optimistic, idealized portrait of the nation's present and future economic health. We felt secure in the knowledge that he would protect and provide for us. We discounted his inattentiveness, dismissed his ineptness, and excused his responsibility for the rampant political corruption that marked his administration in order to preserve his good parent image.

Similarly, for decades we created national enemies by viewing them as unidimensional. We conspired with our national leaders and allowed them to characterize certain other nations as exclusively evil, seldom questioning the one-sided portrayal. But these enemies functioned as the out-group onto which we projected our collective ills. In addition, we maintained our illusion of harmony by sharing a common enemy.

Communism served as our national projection screen for many years. Reagan's portrayal of Russia as the "evil empire" reinforced that illusion. When the Soviet Union collapsed, exposing itself as a paper tiger, we were left without an enemy. Without an adversary to rail against, regressive groups are confronted with their own inner turmoil.

When Iraq invaded Kuwait, we had the ideal scapegoat and enemy in Saddam Hussein. Hussein's own evil nature allowed him to be easily portrayed as the embodiment of all evil, which, of course on the face of it, is absurd. But in regressive groups, where critical thinking is suppressed and dissent silenced, the illusions are easily maintained. Comparisons to Hitler combined with bromides designed for mass consumption—"just versus unjust," "ethical versus unethical," "moral versus immoral"—were used to damn Hussein's evil nature while adding moral weight to our own position. The more evil the enemy, the more virtuous regressive groups feel. And the more virtuous the regressive group feels, the easier it is for it to commit immoral acts.

Just as the Soviet Union served as the out-group and diverted us from our own evil tendencies, now Hussein and Iraq served as scapegoats and projection screens for our individual and collective immorality, destructiveness, and corruption. We condemned Hussein for his invasion of a sovereign country while overlooking our own recent incursions into Panama and Grenada. We denounced him for his excessive military build-up with no acknowledgment of our own. We assailed him for atrocities committed against his own people, but were silent about human rights violations in this country. Many of the negative traits we ascribed to Hussein were mirror descriptions of our own dark side.

The public response to Hussein's demonization demonstrates how regressive groups, under stress, are easily manipulated. Predictably, the country responded with a wave of nationalism—yellow ribbonism and flag waving—that unleashed many of the primitive, impulsive, and destructive aspects of the group: suppression of dissent, suspension of critical thinking, and endorsement of violent behaviors against the out-group.

We depended solely on President Bush's leadership and decision making before and during the war. Dissent was stifled and dismissed as insignificant, even unpatriotic. As we have recently learned, even the President's senior advisors were not willing to voice alternative opinions directly to him.[39] The peace movement, trampled in its nascent stages by the public persecution of dissenters, was virtually ignored by the media. From the commitment of U.S. troops in August 1990, until January 3, 1991, the three major television networks broadcast 3,000 minutes on the nightly news about the Gulf crisis, of which only 29 minutes were devoted to grass roots dissent.[40] Peter Arnett (CNN), who broadcast from Iraq, was savaged as a traitor by government officials because his reporting sometimes conflicted with the official version of the war or its public myth. Independent press coverage was nonexistent.[41] News shows were awash with retired generals, former government officials, and foreign policy analysts all touting some version of the official party line. Once the war started, only one of the 878 on-air expert sources represented a national peace group.[42] Even a paid antiwar advertisement sponsored by the Military Family Support Network, to be aired prior to the start of bombing, was turned down by the networks.[43] However, seven members of the 1991 Super Bowl teams were given a chance to share their "expert" opinions over the air.[44] Although there was some public dissent throughout the war, these minority opinions were not heard.[45]

The Pentagon's manipulation of the press and the media's capitulation fed the national psyche with extremely distorted and sanitized images of war. These pictures maintained the illusion of a "just war." No bloody messes, no body counts, just pictures of the successful smart bombs and Patriot missiles, that reaffirmed the nation's technical superiority. The dehumanizing of Hussein (comparing him to Hitler), his soldiers (falsely accusing them of taking babies from Kuwaiti incubators and leaving them to die; calling them "sand niggers" by U.S. troops), and the Iraqi people by moral exclusion (a "just" war) made excessive brutality possible, even welcomed. Moral exclusion devalues outsiders to the extent that they become less than human, making extreme violence against them acceptable.[46]

As a result of this manipulation, the nation was swept along by a contagion of emotion and patriotic fervor that resulted in support for and indifference toward the dehumanized and barbaric actions against Iraq. The killing of nearly 183,000 people—75,000 soldiers; 3,000 civilians during the air raids; 70,000 civilians in 1991 because of deterioration of health and sanitation conditions caused by the war; 5,000 soldiers and another 30,000 civilians in the civil war immediately following the war; when taken as a percentage of the population, that number equates to approximately 3,000,000 Americans,[47]—the destruction of Iraq's civilian infrastructure, the bomb-shelter massacre of hundreds of women and children, the bulldozing and burying alive of Iraqi soldiers, and the aerial slaughter of soldiers and Asian workers fleeing Kuwait are just several examples of the violence that was perpetrated against the out-group.

The devaluation of Iraq was so effective, and the group's need for that deception so great, that when the facts of the war appeared, there was little public interest in them. So it is safe to assume that a majority of the public still believes in the government's portrayal of a highly successful technological and bloodless war, when in fact fewer than 10% of the 88,500 tons of bombs dropped were "smart" bombs, and 70% of those missed their targets.[48] Only one of the 90 Scud missiles fired at Saudi Arabia was actually destroyed by a Patriot missile, and there is conclusive evidence that the Patriots did extensive damage as they crashed into Israeli streets.[49]

The formidable Iraqi army, estimated at 500,000 troops before the war, turned out to be fewer than 183,000 malnourished soldiers with third-rate equipment.[50] The elite Republican Guard withdrew long before the war began and remains largely intact today.

The bombing campaign, reported to have focused exclusively on military objectives, instead targeted the civilian infrastructure, and 3,000 Iraqi civilians were killed during the air raids.[51] Finally, the notion that Iraq was on the verge of developing a nuclear weapons capacity (which found widespread support among the U.S. public as a reason for going to war) was discredited when evidence that existed before the war reappeared after it, reaffirming that Iraq was at least 3 years away from developing a crude nuclear weapon.[52] The notion that Hussein planned to invade Saudi Arabia was also a lie; intelligence and defense officials knew shortly after the invasion of Kuwait that he had no intention of doing so.[53] A sign that these myths were still intact after the war was President Bush's continued ability to tap into this collective fantasy by using the Saudi Arabia invasion/nuclear bomb myth as he campaigned for reelection in 1992.

The psychological necessity of carrying out these inhuman actions against an out-group rests in trying to purge our own individual and collective evil that we had projected onto Hussein and Iraq. However, it is an impossible task because the evil resides in us and cannot by eliminated by annihilating the enemy.

Since the end of the war with Iraq, the United States has sought to find or create a new enemy. External candidates include Cuba, Japan, Libya, China, or North Korea. Domestically there is a resurgence of racism and anti-Semitism on college campuses,[54] and attacks on political correctness, the "feminist agenda," and affirmative action. Scapegoating efforts directed at minority groups, welfare recipients, and immigrants offer convenient targets that distract us collectively from identifying and reclaiming our own shadow material. Of course, the danger in not creating an out-group is that it might result in the nation having to painfully confront its own dark side!

CHANGING REGRESSIVE GROUPS

Opportunities occur in the lives of regressive groups where a change or transformation is possible. These can develop as impasses in the group, places at which the group is unable to proceed. If the group can focus its attention on the impasse and explore the intrinsic meaning of the "stuckness," an expansion of group consciousness is possible. These moments present opportunities for major turning points in the group. However, attitudes are extremely difficult to change. It requires significant work and often considerable pain. Self-doubt, an openness to self-examination, the willingness to suspend beliefs about what is right and wrong, are necessary for a regressive group to evolve.

In order to change, groups must undergo a period of uncertainty and chaos, a point at which most groups got stuck. In this state where rules are suspended, boundaries ambiguous, and members anxious, the possibility for transformation to a new level of organization is possible. In many ways, this environment is analogous to one aspect of the "creative process," described by artists as a period of intense frustration endured before insight or enlightenment can be achieved.

However, chaotic groups have difficulty tolerating the anxiety and ambiguity inherent in such situations. Oftentimes, a group reestablishes equilibrium or dissipates the energy created by the moment before it has time to reflect on its dilemma. Regressive groups never progress beyond the third stage of development, disunity. This is true of many task oriented groups. Although these groups function adequately and never regress to impulsive or primitive levels, their creative potential remains untapped.

Many regressive groups are incapable of resolving their own predicaments. Once a group has reified its regressive tendencies, change becomes extremely difficult. In some cases, the inflexibility of these groups eventually leads to their dissolution.

Intervention by a trained group consultant is usually the only method of facilitating movement in these groups. Selecting an individual the group trusts and inviting him or her to observe the group and offer feedback can help. However, short of outside intervention, there are some options that members of regressive groups might utilize to effect change. This section examines those strategies and suggests several methods for preventing the growth and activation of the collective shadow.

Contrary opinions could be shared. Group members need to become aware of their potential to influence the group.[55] Silent members often have similar discordant views and can be empowered to share them if others take the lead. However, dissenting members must be prepared to face serious resistance in the form of confusion, denial, and anger. Even scapegoating and expulsion from the group are possible. However, research indicates that dissenters who persist in their views can influence the direction of the group.[56]

Tolerate Ambiguity

In order to change, group members must be prepared to undergo a period of chaos where rules are suspended, boundaries become ambiguous, and members experience anxiety. It requires a committed effort, considerable pain, an openness to self-examination, and a willingness to suspend beliefs about what is right and wrong. This period can be compared to the time of intense frustration endured by artists just before creative insight is achieved.

Unearth Conflict

Conflict must be uncovered and addressed. The group must recognize conflict as a healthy catalyst for change, and norms for its expression and resolution must be developed. Once conflict is openly addressed, members sometimes discover that it was the fear of conflict that was debilitating. Its expression is quite cathartic.[57]

Reclaim the shadow. Another area of reconstruction the group must undergo is the reclaiming and integration of its dark side that was projected onto out-groups. The group must make a conscious effort to acknowledge its negative aspects. A place to start is by examining the characteristics it assigns to competing or enemy groups. Careful examination of those traits often yields hidden information about the group's own behavior.

Furthermore, efforts at developing a more comprehensive and compassionate understanding of the out-group also help illuminate the group's own denied attributes. The more conscious groups become of their negative aspects, the less prone they become to manipulation.

Examine Metaphors

Identifying the source of a group's intractability can be aided by examining the group's metaphorical language. Regressive group metaphors are generated in response to threatening situations. They usually appear in the form of stories introduced by group members and are attempts to resolve the group's dilemma at an analogical level.[58] Even solutions to the group's "stuckness" are often embedded in the metaphor. Several methods for utilizing the information contained in the metaphors were detailed in chapter 8.

PREVENTION

The nature and purpose of a group dictates to a large extent the specific role of the leader and of its membership. Usually, in therapeutic groups, the relationship between the leader and the members is reciprocal, marked by an intensified level of intimacy. In larger, task oriented groups, the relationship is less personal and more unilateral as the leader, for the most part, provides direction for the group. In either case, the freedom to express diverse opinions, air dissenting views, and respect minority beliefs can be fostered by both the leader and group members.

Creating Awareness or Fostering Group Mindfulness

Group mindfulness[59] characterizes an open group wherein exploration of different and diverse opinions is encouraged along with critical questioning that challenges the group's frame of reference. Leaders are initially charged with creating and fostering this tolerant environment. This is best accomplished through allocation of authority by inviting group members to share and participate in the group's decision-making process. In large groups, this can be accomplished by delegating authority to small task groups who are empowered to act on behalf of the whole. All information on which a decision is based should be made available to all group members. Composition of the small group is critical and should reflect the diversity of opinion in the larger group. Group membership in these decision-making groups should be rotated as often as possible.

Promote *Partage*

Group building based on the French word *partage*, meaning both to divide and to share,[60] reflects an attempt to build group cohesiveness on diversity rather than consensus. Underlying this somewhat unorthodox idea is the notion that group consensus promotes conformity "that rules out the other, substituting one set of beliefs for another, brings us the regime of yellow ribbons and American flags as the test of patriotism. It leads students to condemn dissent, as one student did at Princeton years ago, as treasonous and un-American."[61] The notion of partage recognizes that differences are often irreducible, and group rules for coexistence should be developed that do not presume the necessity of consensus. "Partage accepts diversity as a condition of our lives."[62]

Assume Responsibility

Group members must assume personal responsibility for the group's actions and be willing to voice opinions that may run counter to the majority view. Viewing the leader realistically inhibits the development of the omnipotent parent image. Perhaps the most important aspect of personal responsibility is to develop an independent knowledge base of the group's activities. This is particularly true if you belong to a large organization. Sharpen your critical thinking skills by listening to minority positions and reading alternative sources of information about your group. Avoid groupthink! Remain open to the possibility that there are multiple realities. During the Gulf War, it was determined that those people who got the majority of information about the war from the major television networks actually knew less about it than people who obtained their information elsewhere.

Appoint an Ombudsman

Research demonstrates that when groups are continuously exposed to dissenting viewpoints, they are likely to think more actively and creatively.[63] All on-going business, community, and organization groups should consider appointing an ombudsman, or critical evaluator, to monitor their activities and critically assess their behavior. Periodical feedback should then be included as an integral part of the group's activities. Another means to accomplish this goal is to assign, on a rotating basis, the role of critical evaluator to each group member, who then plays the "devil's advocate."[64]

All groups have regressive tendencies that develop from their avoidance of conflict, denial of negative traits, suppression of critical thinking, and damping of dissent. Healthy groups are able to recognize these inclinations and summon the courage to confront them openly. Inflexible groups ignore them, hoping they will just go away. They never do! All regressive group characteristics that are denied or suppressed will continue to haunt the group, in one form or another, until they are acknowledged.

Generative and Transpersonal Groups

Although there is considerable agreement that a collective mind does emerge from the interactions of a group of people, exactly how that mind manifests itself is in question. There are two schools of thought. The first suggests that the group mind assumes an identity greater than the sum of its parts;[1] the second implies that the emergent mind is equal to or less than the aggregate of its parts.[2] The type of group mind that emerges is predicated on the group's level of development. Hence, groups stuck in the forming stages manifest a less developed mind than those groups in more advanced stages. These less developed or regressive groups were examined in the last chapter.

In order to differentiate between groups that mature and those that do not, it is helpful to picture a continuum. At one end of the continuum is the regressive and least developed level of group. At the opposite end is the generative and highest level of group development. This chapter describes the characteristics of mature (generative) groups, and speculates on the attributes of the group mind. The phenomena of the group mind are not discussed in the contemporary literature. There was considerable speculation among earlier theorists, but in recent years little has been written about it. However, the notion of a universal mind or consciousness occupies a central position in transpersonal psychology.[3]

Many theorists from various disciplines[4] have "argued in one form or another that groups, institutions, and culture have a life or existence quite separate from the individuals" who comprise them.[5] This emergent entity is called many names: group mentality,[6] group theme,[7] group metaphor,[8] universal mind,[9] and simply, the group mind.[10]

The group mind has been described as "individual minds, forming groups by mingling and fusing, give birth to a being, psychological if you will but constituting a psychic individuality of a new sort."[11] However, it has also been rejected as "ill-considered and scientifically pretentious psycho-mysticism, quasi-fascist nonsense" and dismissed as "a metaphysical assumption with no scientific justification."[12] Others not only support this idea of a group mind, but advocate that earlier definitions be expanded:[13]

> the unanimous expression of the will of the group, contributed to by the individual in ways of which he [sic] is unaware, influencing him [sic] disagreeably whenever he [sic] thinks or behaves in a manner at variance with the basic assumptions. It is thus a machinery of intercommunication that is designed to ensure that group life is in accordance with the basic assumptions.[14]

One theorist argues that the group mind not only transcends the sum of the individual minds that form it, but each mind is altered as a result of belonging to it.[15]

The group mind concept eventually lost much of its influence because it was too unwieldy and amorphous to be of any use.[16] It fell subject to what Allport[17] described as the *group fallacy*, "the error of substituting the group as a whole as a principle of explanation in place of the individuals in the group."[18] Allport explained that no psychology of groups existed that was not actually a psychology of the individuals who comprised the group. The individuals, he concluded, were the proper locus of explanation:

> One of the recurrent problems faced by pre-Allportian social psychology was that of describing the relationship between the individual and social, of forging a conceptual bridge between the two. Group mind theorists found this a difficult issue. Durkheim viewed the individual as superfluous in sociological analysis. Wundt had no clear answer as to how important phenomena of community life affected individual consciousness, and McDougall was forced to postulate a sympathetic mechanism and social instincts which linked the individual to the social.[19]

Despite the many interpretations by theorists of differing philosophies, a prevailing description emerged of the collective mind as an "autonomous and unified mental life in an assemblage of people bound together by mutual interests."[20] One theorist thought the concept of group mind, although valuable and sound, was limiting. He argued that it could not be viewed as a separate entity, but only as part of a cosmic mind.[21]

UNIVERSAL MIND

Transpersonal psychologists have long championed the idea of a cosmic or universal mind.[22] It is described as a Helping Potential or creative force that manifests itself in "works of the human spirit: heroic or spiritual actions, science, art, and philosophy."[23] It is personified as a welcoming and loving entity with directing capabilities.[24] The universal mind is further portrayed as capable of providing harmony, meaning, and growth, and is associated with the creative force behind the cosmic design:[25]

> Mind at Large resonates with the idea of God as love, eros, agape, karuna and with the idea of a friendly intelligence somehow hidden in the secret byways of a dangerous world. . . . Mind at Large doesn't refer to a thing; it points to a process. That is part of what is meant by saying that Mind is creative.[26]

A further description of mind compares it to that of an evolving god, a god who exists within us, and who can only emerge through us. Moreover, this "Helping Potential lies hidden in the deep grain and texture of being."[27]

Unlike the conventional religious concept of God, which posited a one-way relationship between creator and creature, it is suggested that the cosmic mind is capable of learning and growth through interchange with the individual minds comprising it.[28] This distinction between mind and traditional definitions of God is clarified:

> Mind that reveals itself in the development of life on this planet is clearly not omnipotent, otherwise it would have assembled perfectly designed organisms . . . without having to go through the long process of trial and error which we call evolution.[29]

From the discussion, it appears that individual and group minds are interacting elements of a larger mind. Likewise, the Gaia hypothesis personifies the earth as one living system, composed of smaller, interdependent systems.[30]

INTERMIND COMMUNICATION

The ability to communicate between minds was explored by transpersonal psychologists.[31] One of the earlier theorists' notion of the group mind led him to speculate on the possibilities of "telepathic interaction" among group members.[32] This idea of telepathic communication found support among many of the traditional theories of Extra-Sensory Perception (ESP).[33]

Koestler was interested in this area and likened the group mind to a "subterranean pool," accessible to individual minds through which they can communicate with one another. He notes:

It looks as if telepathically received impressions have some difficulty crossing the threshold and manifesting themselves in consciousness. There seems to be some barrier or repressive mechanism which tends to shut out from consciousness, a barrier which is rather difficult to pass, and they make use of all sorts of devices for overcoming it. Sometimes they use muscular mechanisms of the body and emerge in the form of automatic speech or writing. Sometimes they emerge in the form of dreams, sometimes auditory hallucination. And often they can emerge in a distorted and symbolic form. It is a plausible guess that many of our everyday thoughts and emotions are telepathic or partly telepathic in origin, but are not recognised [sic] to be so because they are so much distorted and mixed with other mental contents in crossing the threshold of consciousness.[34]

The individual mind is connected with the universal mind when each person is viewed as potentially mind-at-large, capable at each moment of "remembering all that has ever happened to him and of perceiving everything that is happening everywhere in the universe."[35] The primary function of the brain and nervous system is seen as eliminative rather than productive. This system functioned to protect the individual from inundation by a mass of irrelevant information. Thus, it allows through only knowledge that is practically useful and necessary for survival.[36]

The individual mind mirrors in miniature the group mind, yet no individual mind . . . can subtend the whole range of awareness of the group mind. Nor can any individual mind, however great its feeling or desire to act, equal the power and intensity and richness of feeling included in the group mind.[37]

Jantsch[38] conceived of intermind communication as the result of a kind of resonance among individual minds. Earlier we discussed this resonance process, known as phase-locking. This commingling of consciousness is also characterized as the link between the social and spiritual spheres.

In his book, *The Final Choice*, Grosso speculated that in times of crisis, when the rational, conscious self reaches an impasse, the potential for receiving help from the Universal Mind is increased. To support his arguments, he drew many examples from psi research, out-of-body experiences, near-death experiences, and helping apparitions.[39]

MIND AS CONSCIOUSNESS

Jung divided mind into four levels: Ego Consciousness, Personal Unconscious, Collective Unconscious, and the Psychoid Level. These levels roughly correspond to the various levels of mind discussed in this chapter. Thus, using Jung's framework, Ego Consciousness and the Personal Unconscious compose the individual mind, the Collective Unconscious reflects the group mind, and the Psychoid Level is equated with the universal mind. There are, of course, conscious aspects to the group and universal minds that are known to their members.[40] However, this chapter is concerned only with that portion of the collective mind that is outside the conscious awareness of its constituent members.

Each level of the unconscious is whole and part of the larger unconscious that subsumes it. An individual's view of self as a separate entity contains the first two levels: Ego Consciousness and the Personal Unconscious. Together they form the conscious and unconscious halves of self. Jung, however, postulated that beneath the Personal Unconscious there existed the Collective Unconscious, "a vast realm wherein our minds are, like the threads in a tapestry, linked with other minds."[41]

The Collective Unconscious links past and present minds, and is the repository of racial and evolutionary memory. Loye metaphorically likens it to DNA, which transports the genetic code from our past that influences our present and future. Likewise, the Collective Unconscious is transmitted, perhaps in a similar fashion, to or through DNA, from generation to generation, bearing with it all of the cumulative past evolutionary experiences that humans share. Similarly, it shapes our present and future lives. Below the Collective Unconscious is the Psychoid Level. It represents the melding of mind and nature: indistinguishable, inseparable, one. It is synonymous with the universal mind:

> The Psychoid Level . . . is open to influences of every possible kind. It is accessible to whatever forces and factors happen to be present at a given moment in the continuum of the Self, whether these are factors operating within one's own psyche, within the psyche of others, or whether they are forces of any other kind active in the universe.[42]

The terms *group mind* and *universal mind* are less cumbersome than the labels *Collective Unconscious* and the *Psychoid Level.* Therefore, they are used almost exclusively throughout this chapter. Remember, these concepts are interchangeable.

THE OMEGA POINT

Teilhard de Chardin's vision of the universe suggested the interrelatedness of all things, particularly the physical world with that of consciousness.[43] His cosmos contained an exterior, material world lined with an interior reality or consciousness. He labeled this inner side of nature the Noosphere or sphere of mind, equating it with the biosphere or organic side of life on earth.[44] Teilhard de Chardin believed that as the external world community developed and moved closer toward centralization, there would be a corresponding commingling of minds that would form a global consciousness or omega point:

> The omega point is to the individual minds which form it as the individual human mind is to the neurons which form the brain. . . . it unifies and centralizes the activities of its constituent minds in a fashion not unlike that in which the activity of the individual human mind draws together and centralizes the activities of the nerve cells of the brain.[45]

The commingling of minds occurs in a manner consistent with the transcendent moments that arise in generative groups, "not through loss of individuality but through a mutual enfolding of the most personal inwardness of each individual with other individuals."[46] The end result in both cases is the experience of love.

MIND AS HELPER

Teilhard de Chardin's notion of the omega point or global consciousness is comparable to Grosso's view of the universal mind as an organizing influence and helping potential in the world.[47] But how does this universal wisdom or cosmic mind communicate with its constituent minds? Or intervene in times of crisis? By what mechanism or mechanisms does mind-at-large make itself known?

Theorists proposed various means by which information is transmitted from or access gained to the universal mind. Dream analysis, guided imagery, meditation, zen, and yoga have all been suggested as means for establishing intermind connections.

The triggering of synchronistic events[48] emanating from the activation of an archetypal theme is also seen as a manifestation of the universal mind, and precognition was proposed as evidence of a higher consciousness.[49]

SURRENDER

In each of these cases, the connection to mind or more encompassing levels of unconsciousness represents an internal rather than an external experience, usually preceded by surrendering or letting go of consciousness. This surrendering process of moving inward and downward corresponds to the commingling of individual minds or consciousness with the Collective Unconsciousness depicted by Jung. It has been likened to a diver descending into the ocean's dark reaches:[50]

The deeper "layers" of the psyche lose their individual uniqueness as they retreat farther and farther into "Lower down," that is to say as they approach the autonomous functional systems, they become increasingly collective until they are universalized and extinguished in the body's materiality, i.e., in chemical substances. The body's carbon is simple carbon. Hence "at the bottom" the psyche is simple "world."[51]

This ability to "surrender fixed and partisan patterns in favor of a more fluid and spontaneous interplay of autonomous yet 'globally aware' persons seems to be a crucial factor associated with the achievement of more subtle levels of consciousness."[52]

The emergent mind, in generative groups, manifests itself as individual members surrender their consciousness to the group. This surrender is a kind of letting go. Trusting in the group's safety, members are able to relinquish their ego boundaries and connect, not only with one another, but with the universal mind. It is important to note here that individuals do not lose their identity to the group, but integrate the "I" with the "We." Through this surrender, group members become spontaneous, rather than autonomous, and are engaged and carried along, often experiencing moments of transcendence.[53]

The word *transcendent* is perhaps a misnomer when used to describe the process by which individuals gain access to the universal mind. In this case, transcendence means descendence. The process by which the ego is surpassed is literally one of turning inward and descending to deeper, more encompassing levels of unconsciousness or mind. The same is true for the collective process that occurs in generative groups, until ultimately the Psychoid Level or omega point is reached, where mind and matter are indistinguishable. Here everything is pure energy; past, present, and future states coexist. At "these moments of isomorphic integration the individual gains access to a vastness of possibilities which is of cosmic proportions."[54] These moments are described as "being in harmony with nature . . . this ability to yield, to be receptive to, or respond to, to live with extrapsychic reality as if one belonged with it, or were in harmony with it."[55] One means by which psychic energy, released or tapped at the Psychoid Level, takes form is the archetype.

ARCHETYPES

Jung assumed the existence, within the collective unconscious, of an amorphous energy force that is part of the inherited structure of the psyche. These thought forms or *archetypes*, as Jung called them, are based on ubiquitous motifs or themes that, because of our common evolution, transcend racial and cultural boundaries. These themes have existed throughout the evolution of humankind.

Jung emphasized that the content of archetypes is not predetermined. They represent dispositions or forms that "draw the stuff of experience into their shape, presenting themselves in facts, rather than presenting facts."[56] In other words, the content of these primordial images becomes manifest when "filled out by the material of conscious experience." So although archetypes are universal themes for humankind, their manifestations or content are unique, each given text from an individual's personal experiences.

Jung's major archetypes include aspects of the personality: the persona, the shadow, the anima and animus, and the self. Other common archetypal images include the wise old man, the hero, the great mother, the Earth Mother, and the divine child. In addition to these images, there are a multitude of archetypal themes, often expressed through fairy tales, folklore, or religious or mythological motifs, that influence our lives: death and rebirth, good and evil, devouring monster, the divine child, the eternal return, magic effect, paradise and the fall, transformation, and so forth. Archetypes are not necessarily discrete and oftentimes overlap, so it is very difficult to quantify them.

Archetypes represent predispositions to respond to certain experiences in specific ways.[57] They are often activated in times of stress or crisis, and are experienced as a "sense of transcendent validity, authenticity, and essential divinity."[58]

> For the individual caught up by the power of an archetype, the symbolic meaning of his or her life is no abstraction but a powerfully felt and utterly convincing reality which, with or without the individual's conscious participation, directs and forms the nature of the world.[59]

This directing or helping potential of the archetype seems akin to Grosso's notion of the universal mind. When describing the properties of helping apparitions, as one manifestation of the universal mind, Grosso concludes that their appearance occurs during times of transition, crisis, and danger:

> They arise, for instance, when the rational, conscious self faces an impasse; we might say that helping apparitions tend to arise when reason can no longer come up with a rule for coping with a crisis. A willingness to admit helplessness seems to clear the way for the helping function.[60]

The external appearance of a helping apparition may be an internal manifestation. Possibly during times of crisis, reflection of an inner self is hallucinated outward in order to bring forth knowledge or wisdom buried in the collective unconscious. On these occasions, activation of an archetypal image or theme is projected outward in the form of a helping apparition:[61]

> In Jung's thinking, the activation or awakening of an archetype releases a great deal of power, analogous to splitting the atom. This power, in the immediate vicinity of the psychoid process from which the archetype takes its origin, is the catalyst for the synchronistic event. . . . The idea is that the activation of an archetype releases patterning forces that can restructure events both in the psyche and in the external world. . . . The power that is released is felt as numinosity—literally a sense of the divine or cosmic.[62]

These forces manifest themselves as synchronistic events, and in groups, through symbolic expression as generative metaphors. In both instances, they promote understanding, offer resolution to the crisis, and suggest possible future outcomes. In generative groups, the synchronistic events and the generative metaphor are organized around a similar theme. "Meaningful coincidences frequently involve several events which, though radically different in form—one may be an idea, another a physical object—are tied together by a common pattern or theme."[63]

Synchronistic Events

An often cited example of synchronicity, linking the activation of an archetype in the psychic with an outer event, comes from Jung:

> A young woman I was treating had, at a critical moment, a dream in which she was given a golden scarab. While she was telling me this dream I sat with my back to the closed window. Suddenly, I heard a noise behind me, like a gentle tapping. I turned around and saw a flying insect knocking against the window-pane from outside. I opened the window and caught the creature in the air as it flew in. It was the nearest analogy to a golden scarab that one finds in our latitudes, a scarabaeid beetle, the common rose-chafer, which contrary to its usual habits has evidently felt an urge to get into a dark room at this particular moment.[64]

The circumstances, at a critical moment in analysis, an impasse, may have activated this archetypal image, thus triggering the synchronistic event. This meaningful event was the turning point in the woman's therapy. As Jung points out, the scarab motif is a symbol of rebirth, and subsequently the woman underwent a renewal.[65] The triggering of synchronistic events is rooted in the Psychoid Level where both inner and outer reality, past and future possibilities, and psychic and physical worlds coexist.[66]

Synchronistic events present a very common learning opportunity experienced by most people: the simultaneous appearance of two seemingly unrelated events. Assume for a moment that the name of a friend who lives in a distant city enters your mind. You ruminate about that friend for a few minutes, perhaps puzzled by what brought the name to your attention. Suddenly the phone rings and it is she. Usually it is dismissed as a chance occurrence. But what if you regard it as a message from the collective unconscious, a message that signals to you a learning opportunity, symbolically connected to your friend?

This learning opportunity is similar to what occurs in generative groups through the development of a collective metaphor. In this case each member is simultaneously connected with the collective theme, although the learning potential is different for each member.

Generative Metaphors

Archetypes are expressed to individuals through their dreams and fantasies. In groups, Jung notes, the archetype is transmitted through myth and, by extension, metaphor.

The function of metaphorical development in groups was examined in chapter 8. Primarily, group metaphors in the forming stages of groups provide members with relief from excessive anxiety by creating a symbolic safety valve through which they can communicate their concerns about group issues deemed too unsafe to discuss openly. What role, if any, does metaphor play in the advanced stages—where conflict was successfully resolved; where high levels of safety and trust exist; where strong cohesive bonds were established; where members can transcend their own ego needs in the service of others; where cooperation is the norm rather than the exception?

The development of metaphors in highly cohesive groups may serve as a communications link between the collective unconscious and the group unconscious. Messages from the Psychoid Level or universal mind may be communicated through metaphors to the group, whereby group members are able to tap into a collective wisdom and experience perceptions beyond the reach of their ordinary senses: healing may occur, and the future may be glimpsed. "In short, many of the phenomena referred to as 'psychic' or 'transpersonal' may appear."[67]

In keeping with the nomenclature used to describe highly cohesive groups, and as a means of distinguishing between previous discussions of metaphors and those in this chapter, the term *generative metaphor* is used. The generative metaphor, like Jung's symbols, can be seen as a means of transmitting wisdom from one level of unconscious to another:

> The unconscious speaks to the conscious in symbols and analogies; the collective unconscious, the ancient storehouse of the accumulated wisdom of the human race, speaks to the personal consciousness in stories and parables; the Higher Self speaks to the ego-personality self in the language

of myth and metaphor. This great spirit, or Atman, has access to the buried strata of our psyche, where it can awaken the sleeping memories of long past experiences; and being beyond time, it can send us messages from the future. . . .[68]

Hence, generative metaphors represent a shared consciousness among group members and with the larger universal mind, and offer extraordinary opportunities for the group and its members. One description of the process by which fantasy themes are generated in groups seems comparable with how generative metaphors are developed.[69] Furthermore, it is suggestive of a larger consciousness at work; when a psychodramatic image presented by one person "is picked up and elaborated by others, in spite of ample opportunity for other images to have been presented (and for the first image to have dropped out of attention) then one may suggest that there is some motivational incentive in the selection and elaboration that is not confined to the original presenter of the image, and that those persons who continue to elaborate the fantasy theme are somehow reinforcing each other."[70]

Metaphors, like myths, signal possibility: a potential for learning, insight, and discovery; an invitation to turn inward and uncover the personal meaning contained within the symbols. Campbell, like Jung, has written compellingly about the potential of myths. He contends that "Myths are clues to the spiritual potentialities of the human life."[71] ". . . Myth is a manifestation in symbolic images, in metaphorical images, of the energies of the organs of the body."[72] ". . . Myths and dreams come from the same place. They come from realizations of some kind that have then to find expression in symbolic form."[73] The language of experience is discussed as the language of myth:

> . . . a language that bridges the space between the conscious and unconscious, a language that speaks of the meaningfulness of events both within the mind and within the greater world in which each of us lives our day-to-day lives. Myth draws us into meaningful relationships with our entire world, an arena that encompasses both our minds within and the objective events of the world without.[74]

The generative group metaphor, too, serves as a bridge between the larger mind and its constituent minds. Not only does the metaphor bear symbolic messages, it also acts as a harbinger that allows group members glimpses into the future. The generative metaphor, in a sense, facilitates precognition. However, the future that is glimpsed is not fixed, but exists as possibilities or potentialities or as suspended alternative states.[75] To better understand this notion of precognition and provide an overall framework for this thesis, we turn to Loye's model of the holographic universe.

THE HOLOGRAPHIC WORLD

Loye presents two different views of a holographic universe: one vertical and predetermined; the other horizontal, encompassing free will. The former is based on Bohm's world view; the latter represents an extension of that theory developed by Loye. Loye's models draw on the works of psychologist Jung, neuropsychologist Lashley, neurosurgeon Pribram, and physicist Bohm. Loye creatively weaves their ideas of the collective unconscious, synchronicity, the hologram, and implicate and explicate orders, with the notions of free will and determinism. Before elaborating on Loye's theories, it is first necessary to understand the hologram. Bohm and his mentor, Albert Einstein, were both motivated by the ethical implications of viewing the world as fragmented and disconnected. Both were impassioned with the idea of discovering a unified field theory that would demonstrate the underlying connection of everything in the universe and solve the problem of order.[76] But unlike Einstein, who placed the human observer outside of his or her objective universe, Bohm sought to eliminate the subject–object dichotomy by visualizing the universe as a single, unbroken whole containing both "thought (consciousness) and external reality as we experience it."[77]

Born of his interest in the lens and its role in science of "providing images of things that couldn't otherwise be seen,"[78] Bohm posed the question, "Is there an instrument that can help give a certain immediate perceptual insight into what can be meant by undivided wholeness, as the lens did for what can be meant by analysis of a system into parts?"[79]

The answer was the hologram, an image produced by splitting a beam of light so that a whole image is reproduced, rather than the usual "flat surface illusion."[80] What was particularly intriguing about the hologram was the way in which information was spread throughout the image as a whole rather than in parts. In other words, each part of the hologram contains the whole, although the clarity of the whole diminishes as light is projected through smaller and smaller fragments.

Pribram, building on the early works of Lashley, Gabor, and van Heerden, applied the holographic idea to his brain research. In doing so, he was finally able, among others things, to postulate how memory was stored throughout the brain.

Bohm began collaborating with Pribram to describe a holographic theory of the universe that showed how "separate objects could in fact be connected together in the underlying, interpenetrating way that physics already indicated was operating in the universe."[81] For Bohm, the hologram suggested an interacting relationship between the seen or explicate order, and the unseen or implicate order.

The implicate order, or nonmanifest reality, and the explicate order, or manifest reality, exist side by side. Loye borrowed Alice's Wonderland to help us understand the two interconnected realities. From our side of the looking glass, we would "be aware of ourselves and everything in this 'real' world of ours as being solid kinds of things separated by space and living according to the ticks of time."[82] Peering from our world into the implicate order (Wonderland), "All we can see is a big blur to which bubble-chamber photographs and the theories of physics give us some slight clues."[83]

However, Loye speculated that this would look very different if we observed ourselves from the Wonderland side of the looking glass. Then we would find ourselves in a "superrich reality in which everything was interconnected in a 'ball' of spacelessness and timelessness."[84] Looking to the other side, we would see "what appear to be a number of awfully funny shapes and sounds produced by an interface between our worlds of holographic waves, which come racing into and out of being within every other moment, much as pictures are projected on a movie screen."[85] In Wonderland, the transmission of information is no problem because everything is contained or embedded within everything else. But, in the "movie world," the "sights" and "sounds" must cross barriers of "empty space" or "time."

Loye next asked the questions, "How is the gap bridged between the two worlds?" and "How do we move from one ordered universe into one of spacelessness and timelessness?" Bohm speculated that the vehicle is "insight." As we are part of the implicate and explicate order, so our consciousness is connected with both sides. Although the mechanism is unknown, Bohm's theory at least provided a rationale for the association between the two worlds.[86]

THE VERTICAL HOLOGRAPHIC ORDER

Loye labeled Bohm's paradigm the *vertical holographic order*. It is similar to the self-organizing, hierarchical universe described in the theories of Prigogine, Jantsch, and von Bertalanffy. Each theorist depicted a hierarchical world starting at the smallest level and moving upward to ever-increasing levels of complexity (e.g. the organs, the body, the individual, the family, etc.), with each level self-contained yet forming the building blocks for the next level of organization. Each level operates according to its own internal laws while participating in a greater whole.[87] Perhaps the best depiction of this multilayered universe comes from Koestler:

> The living organism and the body social are not assemblies of elementary bits; they are multi-levelled, hierarchically organized systems of sub-wholes

containing sub-wholes of lower order, like Chinese boxes. The sub-wholes—
"holons," as I have proposed to call them—are Janus-faced entities which
display both the independent properties of wholes and the dependent properties of parts.[88]

This notion of a vertical universe leads to the inevitable conclusion that
everything is predetermined, existing within a single giant hologram.[89] The
future, in this state, like many of those in Eastern religion, is fixed, and
precognition is a matter of discovering it.

THE HORIZONTAL HOLOGRAPHIC UNIVERSE

Loye proposed a rather elegant alternative paradigm incorporating free
will: the horizontal holographic universe. He speculated that each of us is
individually and socially bound with those meaningfully connected to us
within, which from a "celestial eye" would appear to be large, amoeba-like,
holographic entities that exist side by side with similar entities. Within
these entities, or interholographically, everything is predetermined for a
time. By glimpsing the implicate order within the hologram, one could
foresee the predetermined future. Then, he proposed, as these restless,
shifting amoebas move, they bump into one another, engulfing "small bits
of substance," and occasionally one holographic entity completely swallows
another one. We seldom notice the amalgamation of small bits of substance
that produce microscopic changes, but as Loye suggested, we are "jolted"
by the engulfment of our entire entity or by our own absorption of other
whole entities, which causes the macroscopic changes in our lives. Inter-
holographic conversion would be akin to first-order change; changes that
occur as a result of engulfment seem similar to notions of second-order
change. Precognition in this second situation occurs intraholographically.
As Loye explained:

> Precognition would act as a leap (hololeap) across the gap between the
> jostling holograms, A and B. Or at the first contact it might, with some
> heightened burst of excitement, range out from someone in A throughout
> the engulfing mass of B. The purpose would be to provide a reading as to
> the personally or socially meaningful contents of B, the engulfing mass, to
> this person A who would be acting as both seer and engulfee.[90]

In this second scenario, free will is engaged and reality is created, all
". . . set in motion by aspiration, by curiosity, by desire, by boldness, openness, and courage—by that belief that it can be done, which paranormal
research has shown is such a constant in successful telepathy and healing—might lie the opportunity both for the seer to perceive and for the
activist to influence the shaping of futures that are not predetermined."[91]

Thus, as we bump into other amoeba-like entities, we may catch glimpses of possible alternative futures that each exist in a present suspended state.[92] Additional support for this view comes from the quantum physicist Schrödinger and his hypothetical cat. The cat is placed in a sealed box with a device that, if triggered, will kill it. An electron is introduced into the box whose action may or may not trigger the device. Prior to opening the box, the question is asked, "Is the cat dead or alive?" Of course we do not know the answer, but one of two outcomes is assured—the cat is either dead or alive. But viewing the situation from a quantum perspective, the cat is neither dead nor alive. Both possibilities exist side by side until the box is opened and one becomes manifest.[93]

Sometimes engulfment is random, reminiscent of the existentialist Sartre's thrown condition, for which there is neither explanation nor cause. However, there may be occasions when the particular entities that we engulf, or are engulfed by, are future conditions that have been influenced or created, as Loye suggested, by our curiosity, desires, wants, and needs. In these cases, we participated in the creation of our future condition.

MIND OR CONSCIOUSNESS

In our vertical, multilayered universe, the relationship among individual minds or consciousness, group minds, and the universal mind is easily understood. Each holon is a self-contained, self-organizing, mindful unit. Within each level mind is a whole, and simultaneously a part of a more complex, more encompassing level of mind. For example, the individual mind is both whole and part of a larger group mind that in turn is subsumed by the universal mind:

> If we are each parts of a larger whole—that is, if our minds and bodies are, in effect, holograms within the larger hologram of the universe—then there is no transmission problem, for the information is already within us![94]

Huxley described each person as potentially mind-at-large, capable at each moment of "remembering all that has ever happened to him and of perceiving everything that is happening everywhere in the universe."[95] He contended that the primary function of the brain and nervous system was eliminative rather than productive. This system, he suggested, functioned to protect the individual from inundation by a mass of irrelevant information. Thus, the brain and nervous system allowed through only knowledge that was practically useful and necessary for survival.

In the horizontal universe, the relationship between individual and group minds or consciousness can be viewed similarly to that in the vertical world: Individual minds within each amoeba-like structure collectively form a group mind. But what about the universal mind, or Psychoid Level? Where does it reside? I suspect that it exists within and without each amoeba-like structure. Inside each entity, the individual minds and the group mind they produce are both pieces of the larger universal mind. Like the holograph, the universal mind is contained within other minds. Outside, the larger mind creates the atmosphere through which each amoeba-like entity moves.

These amoeba-like entities literally swim in the larger mind, or in Teilhard de Chardin's words, "the global consciousness." The relationship between the constituent minds and the larger mind are similar in both Loye's and Bohm's hypothetical worlds, with but one exception. In the vertical world, the universal mind is totally contained. In the horizontal world, the larger mind is both part of and contains the world in which it exists.

The horizontal world fits comfortably, too, with the notions of Bohm's implicate and explicate order, and Teilhard de Chardin's exterior material reality lined by an interior reality or consciousness. Inside each holographic entity exists the conscious, manifest, physical world. The outside lining to these entities, in which it swims, consists of nonmanifest reality, or the Noosphere. "All physical phenomena—such as sunlight, molecules, trees, and stars have two parts, a field aspect and an energy aspect."[96] These halves correspond to the inner and outer parts discussed previously.

Returning to our holographic metaphor, individual and group minds are elements that compose the universal mind, and as such, each element literally contains the whole, albeit less defined, less clear, and perhaps buried deeply in the individual's or group's unconscious. In both the vertical and horizontal worlds, the function of mind is similar, and the means by which individual minds, group minds, and the universal mind communicate are identical.

Changes in the horizontal holographic world sometimes occur spontaneously, are triggered sometimes by events inside the entity, and sometimes by events outside the entity. The Noosphere or Mind-at-Large occasionally intervenes in the process of global evolution. Sometimes entities and their constituent members influence their own futures by their wants, needs, and desires; occasionally collision of entities is random.

Summary

When a transpersonal event is experienced within a generative group, it is often precipitated by the need or needs of one or more of the group's members. The context for that event is the group, in which each member participates in creating synergistic conditions that make it possible. On occasion, it is likely that the entire group shares in the transpersonal experience. Individually or collectively, when the Psychoid Level is accessed, future conditions can be influenced. Activation of an archetypal theme, contained in the larger mind pool, represents the disposition or form the future event may take. Manifestations of the themes in the group occur psychically through synchronistic events and symbolically through the generative metaphor. These manifestations provide the bridge by which future conditions are glimpsed.

GENERATIVE GROUPS

These highly evolved and cohesive collectives, where phase-locked members can realize their connection with a greater mind, are known as *generative groups*.[97] Generative groups are highly conscious collectives that have reached the Harmony or Performing stage. Members in these groups are no longer preoccupied with safety needs, boundaries, or roles. Norms of behavior have been negotiated, and the expression of conflict and methods for resolving it have been established. These groups are highly productive and capable of producing second-order change in group members.[98] There is a sense in these groups that all members are part of something that is quite literally bigger than they are and that connects them with the accumulated wisdom of humankind.

Generative groups consists of individuals who are altruistic, trusting, responsible, cooperative, and peaceful. Group characteristics are goodwill, respect, humility, curiosity,[99] love, trust, and a willingness to open to the outer world, relinquishing the need for power, symbol of the ego.[100] Collectively these individuals are capable of a kind of group transcendence in which each member transcends his or her own self-interest for the good of the group.

At this level of development, the group mind is more than merely additive. This mind is analogous to the Higher Power that emerges from the cooperative and altruistic interactions of members of addiction treatment groups such as Alcoholics Anonymous.

John Phillips, in his autobiography of the Mamas and the Papas, a 1960s rock band, characterized this emergent quality of generative groups when describing a rare moment when the band was performing and all four of its members were singing in harmony. He related, "It was almost as if you could hear a fifth voice." This glimpse into the potential of generative groups suggests the emergence of a collective spirit or mind.[101] Halling et al. depicted this emergent mind in a generative group of six researchers who met over time to discuss and synthesize the research data they had collected:

> . . . through dialogue with each other, our interpretations were enlarged. It
> is clear that the interpretation is not simply the aggregate of the six people's

ideas. The scope of the understanding and the direction that emerged, while obviously dependent on the six individual perspectives was not merely additive: like a tapestry, the interweaving of perspectives led to a richer and fuller picture than six individual threads. . . . As a result we were never stuck for any length of time, the level of energy and excitement remained high, no one felt unduly burdened, and we were able to move through the process more quickly than if this had been an individual endeavor. In a real sense the project seemed to have a life of its own and was a support for us rather than a chore.[102]

Whitaker and Napier provided an example of a generative family group and its emergent mind. A family member described one particular therapy session:

But there was also something—a kind of electricity—going on between us Brices. Always. And it always felt intense, as though something important were at stake. What struck me that day was that this process that was happening between us was bigger than all of us, that it had a life of its own. I remember the moment so clearly, sensing the power in the room and feeling a little anxious in the face of it.[103]

Finally, in even describing the relationship artists develop with their work, we can see this emergent force. The painting having "a life of its own . . ."[104] can involve artists such that they find themselves "in a special state where the work seems to be doing itself through us. We are merely the medium, the empty tunnel for the connection between a powerful source and the work in progress and we feel very humble and very powerful at the same time. . . . We become connected, energized, or 'in the rhythm.'"[105]

GENERATIVE GROUP MIND

In generative groups, individuals are capable of achieving integration with other group members and with the collective mind. "The collective mind is also integrated within itself and with the larger world of which it is a part. In these moments of isomorphic integration, the individual gains access to a vastness of possibilities which is of cosmic proportions."[106] These transcendent moments are described as "being in harmony with nature . . . this ability to yield, to be receptive to, or respond to, to live with extrapsychic reality as if one belonged with it, or were in harmony with it."[107] In these moments of connection, group members can sense the magnitude and power of this collective force. They often describe it as overwhelming, frightening, omnipotent, and as having a life of its own.

Mind, described here, is synonymous with consciousness. Individual consciousness and group consciousness are but aspects of the universal consciousness. The mind or spirit that emerges in generative groups occurs when members phase-lock. The fusion of individual consciousnesses creates what has been described as a group mind. However, it does not exist "out there." It is an internal connection sensed in the body by each member and experienced as a moment of transcendence, individually and collectively:

> In these groups there is an experience of oneness, where individual and group consciousness become unified. . . . There is a true meeting, an I–Thou encounter between group members. In this state of oneness dreams can be shared, people may experience perceptions beyond the realm of ordinary senses, healing often occurs, the future is sometimes glimpsed. In short, many of the phenomena referred to as "psychic" or "transpersonal" often appear.[108]

Group members do not lose their identity to the group when individual consciousnesses combine, but integrate the "I" with the "We." Group members become spontaneous, rather than autonomous, and are engaged and carried along, not in spite of themselves, but beyond themselves.[109] Group members do not lose their identity, but integrate each identity within the larger whole. As a result, the group actualizes its intrinsic learning potential. It becomes wise.[110]

THE GENERATIVE GROUP: A CASE STUDY

To highlight the collective potential of generative groups, I have selected a fascinating study conducted some years ago by members of the Toronto Society of Psychic Research.[111] There are other more temporal examples, but this clearly demonstrates the potential of a generative group mind.

In a group psychological experiment, eight people met for 2 hours per week for 1 year and attempted to conjure up a ghostly apparition of a mythical personage they called Phillip. The members had no prior psychic ability and described themselves as ordinary people. During the initial meetings they meditated on Phillip and embellished a fictitious storyline created by one of the members.

A product of these weekly meetings was the strong feeling of rapport and group cohesiveness that developed among the members. Members characterized their relationships as open and trusting. An atmosphere developed that fostered the expression of very personal feelings. Above all, members felt the freedom to give voice to any opinion. Participation in the experiment gave members a sense of well-being, and a greater tolerance and sensitivity toward other people and their feelings. Although the members worked all day and met for 2 hours in the evening for what, at times, was boring work, they expressed a sense of empowerment and exhilaration as a result of their participation in the group, and in no way felt drained by the experience.

One member described an essential characteristic of generative groups when he related that the desire was always for the group's success rather than any individual achievement that might be gained by association with this project. One characteristic that indicated the enormous trust and safety that existed within the group was the childlike atmosphere that was created in many of the group meetings. Members were able to suspend their adult behaviors and recapture the creativity, curiosity, and wonderment of childhood. This childlike creativity was probably instrumental in the ultimate success of the project.

Eventually the group achieved its goal. A manifestation of Phillip appeared by means of a table in the room through which Phillip communicated with rapping sounds. In addition, the table, on numerous occasions, moved throughout the room. For the skeptics among you, this experiment was documented by observers on live television, and was recorded on film, which is available for viewing. Further discussion of the experiment is beyond the scope of this chapter, but the interested reader is referred to Owen and Sparrow's (1976) book, *Conjuring up Phillip*.

The experimenters concluded that Phillip was solely a creation of the group-as-a-whole, the group mind; an expression of their shared subconscious and conscious thoughts, wishes, feelings, and emotions.[112] Phillip was not a disembodied spirit. The responses to the questions asked of him were the responses that the members expected to hear. His personality was a creation of their own minds. Although at times Phillip's responses differed from the outward views expressed by members, closer examination revealed that they were consistent with individual members' subconscious views.

The group factors that contributed to the success of this project, characteristic of generative groups, were complete rapport among the members, an air of harmony, a common motivation and goal, a sense of heightened expectancy, tolerance of differences, open-mindedness, a lack of prejudice, and a childlike quality that created an atmosphere where anything was possible.

For transformative acts to occur, an essential condition that must be established in generative groups is the absolute sense of safety that must be experienced by group members. When members feel safe, completely relaxed, then and only then can they relinquish self-interest and work in concert to achieve a common goal.

However, the self-surrender described here is very different from the mindless resignation and abdication of autonomy that occurs in regressive groups. Generative groups nourish autonomy and individual expression; the surrendering experience is to a larger entity. The ability to "surrender fixed individual and partisan patterns in favor of a more fluid and spontaneous interplay of autonomous yet 'globally aware' persons seems to be a crucial factor associated with the achievement of more subtle levels of consciousness."[113]

This chapter rekindled the discussion of the group mind and sought to bring clarity to it. All groups develop a collective mind, but the manifestation of that mind is dependent on the group's level of development. At the least developed end are psychologically immature or regressive groups. In most cases, these groups are psychologically, emotionally, and spiritually limiting. The inherent potential of these groups and their members remains untapped. At the other end of the continuum are generative groups. The mind that emerges from these groups transcends the individual abilities of their members. This manifestation of mind has direct access to the Universal Mind through which the group and its members can be empowered to extraordinary acts of courage, creativity, and compassion. The therapeutic potential of groups can be maximized by utilizing the transpersonal realm.

Teaching Group Dynamics

The oft heard accusation that the training group is an "artificial" one is not only absurd but ironic, for in fact it is a rather precise analogue of life itself. What differentiates training groups from "natural" task groups is their mortality, their confusion, and their leadership structure. Most groups formed to accomplish some purpose are potentially immortal, have a clear agenda, and a clearly defined leadership. Training groups are born knowing they must die; they do not know, aside from some ill-formulated notions about self-understanding, growth, and knowledge of group processes, why they are there or what they are going to do; and struggle perpetually with the fact that the object whom they fantasy to be powerful and omniscient in fact does nothing, fails to protect them or tell them what to do, and hardly seems to be there at all. Is this unlife like? Is the most persistent theme of the training group situation, the plaintive, "what are we supposed to be doing; what is the purpose and meaning of it all?" a query never heard outside of the esoteric confines of an unnatural "laboratory" setting? On the contrary, it issued from the central dilemma of life itself—that which human beings have always been most unable to face, taking refuge instead in collective fantasies of a planned and preordained universe, or in the artificial imperatives of a daily routine and personal or institutional obligation.[1]

Group experiences provide the best learning opportunities for acquiring group leadership skills. Most introductory group courses offer an experiential component; however, such groups have been the subject of considerable debate. Two ethical questions are raised: "Should students be required to participate in a group experience?" and "Is it possible for students to participate freely in a group experience that is led by the course instructor?" There are no easy answers to these questions in this age of hypersensitivity.

In order to learn about groups, students should participate in them. There is no way this experience can be replicated through lectures, viewing video tapes of groups, or role playing. Given that group experiences are an essential part of the training of group leaders, how can that experience be maximized without compromising either the student or the instructor? Trotzer offers several strategies for our consideration.[2]

1. *Instructor as Group Leader.* The advantage to this model is that the instructor can demonstrate the integration of cognitive knowledge and leadership style. Furthermore, he or she demonstrates a willingness to be seen in an experiential setting. Students are afforded too few opportunities to observe their instructors in action. The instructor can insure that the experience is well managed, safe, and hopefully exposes the student to the entire group developmental cycle. In addition, the instructor and students experience the same event, which makes classroom processing of the experience a significant learning opportunity.

The major disadvantage of this approach is the dual role of the instructor/leader. Students may have difficulty participating freely in such a group because the instructor/leader is responsible for their course evaluation. Furthermore, it is likely that students will encounter the instructor in other courses in their graduate programs. Leading groups requires a high level of emotional participation, and the instructor/leader may have difficulty adjusting to the demands of both positions. Finally, the instructor may have poor group leadership skills.

2. *Outside person as Group Leader.* The advantage to this approach is that it provides students with an independent leader, enabling them to participate freely in the group experience. Choosing a facilitator with excellent leadership skills is of critical importance.

The disadvantage with this approach is that the instructor is unable to effectively coordinate the group's progress with the classroom instruction, so sometimes classroom presentations and group experiences may be at variance. Furthermore, many graduate training programs do not have additional funding to hire outside group leaders.

3. *Outside Leader, Instructor Observer.* With this arrangement, the group is led by an independent leader and observed, usually through a one-way mirror, by the course instructor. However, the disadvantages cited in the first strategy are also applicable here. Students may still be inhibited by observation from the classroom instructor. The problem of funding for the facilitator remains an issue, too.

4. *Leaderless Group, Instructor Observer.* In this situation there is no assigned leader, so the group shares leadership responsibilities while the

instructor acts as an observer. This format gives students an opportunity to examine their own leadership styles. The instructor can coordinate classroom activities with group experience by helping design the format and by setting the agenda for these groups. Further, the instructor is available to process the sessions in class.

However, leaderless groups, comprised of first-time participants, lack direction and leave participants with an inferior group experience. In such models, the instructor often spends a considerable amount of time processing the session. The effects of the instructor's direct observation have already been mentioned.

PERSONAL EXPERIENCE

During my graduate studies, I was exposed to various teaching models. The first, and fortunately for me the best, was an introductory group course team-taught by two faculty members, John Cox and Len Borsari of Springfield College, the institution where Lewin had worked for a short period of time. The course met for 4 hours, 1 night a week, for 10 weeks. Each class was divided into four segments.

The first segment was a minilecture that covered our reading assignments. The second part of the class was a skill-building activity (reflective listening, paraphrasing, etc.) or an exercise related to some facet of group dynamics (group problem solving, conflict resolution, giving feedback, etc.). The third segment was a group experience cofacilitated by the instructors. The last portion was devoted to processing the group session: integrating the lecture, readings, and exercises, with the group experience.

The two professors were exceptionally gifted, able to effectively separate their leader/instructor roles. I never felt inhibited by their presence, nor did I feel that my actions in group would adversely affect my final evaluation. It felt very safe to me, due in large part to the competence of these two individuals.

During my second year, I had the opportunity to colead several of these groups, and my understanding of group dynamics leaped forward. It gave me a chance to try out my own skills under the watchful guidance and safety of an experienced mentor. Each week, in supervisory meetings, I was able to process my work with them.

It was a unique learning opportunity for me, one that gave me a deep understanding and appreciation for the power of groups. It was also the catalyst for me to begin therapy, as I quickly realized how my own issues might impact the group.

My next graduate encounter with groups was also an excellent learning opportunity, but in a very different and negative way. The instructor formed the group from in-class volunteers. It was cofacilitated by an advanced graduate and met over the course of the semester. It was observed through a one-way mirror by the remaining classroom students.

The class was divided into two segments, group and process. During the first section, we observed the group; during the second portion, we processed the group's development in the classroom. Again, I greatly appreciated the instructor's willingness to demonstrate her leadership style. However, the group was a horrifying experience. The instructor was very defensive and remained aloof during the group sessions. It was clearly evident that her own issues were influencing the group. She spent much of the time projecting those issues onto group members and then would eviscerate them for their unwillingness to own those issues. Her coleader was intimidated by her and frequently joined in her attacks. Attempts were made to raise these issues in the process sessions that followed, but they were discounted by the instructor/leader, and quickly the sessions were rendered useless. The instructor refused to look at her role in the group, and the class was a very destructive experience.

It was extremely distressing to watch those students be so violated week after week in the fishbowl, and to feel helpless to intervene. To do so would have been tantamount to academic suicide, so I cowardly chose to hold my breath throughout the course.

However, it left an indelible impression with me confirming that the therapeutic energy of groups can be distorted into a destructive force. I witnessed first-hand how blind spots in the group leader's personality can block a group's growth. Moreover, I observed the power of projection (countertransference) that created double-binding situations from which group members could not escape. I was thankful that my first experience was a good one, because many of my classmates left the course with a distorted view of groups.

The difference between the two courses was the instructors. The first were open to self-examination and modeled that for the class. In the latter example, the instructor was extremely defensive, and should have taught the course in a different manner.

MY TEACHING EXPERIENCES

Since I began teaching 11 years ago, I have taught many group courses and have experimented with many models. I'd like to say that from the beginning I knew exactly what I was doing, but I did not. I learned my capabilities as an instructor/leader through trial and error.

I was committed to the notion of having a strong experiential component in my classes, believing that real learning took place in a group environment. Little did I anticipate the difficulties I would encounter. Throughout these years, I have worked to find an experience that was both safe for my students and that permitted me some access to the group.

My first class, the initial week of a new academic job, was Introduction to Group Counseling. I was excited and full of anticipation for the adventure we were about to embark on. The class was held on Wednesday nights from 5 p.m. to 8 p.m. The course was divided into three segments: a lecture component, a group experience, and a classroom processing session.

I arrived to find 40 students enrolled in the class. That should have been my first clue that my course design was in trouble. However, I allowed my enthusiasm or hubris to overrule my own good sense. I asked for eight volunteers who would participate in a fishbowl experience with me throughout the 10-week quarter. The remainder of the class would observe the group through the one-way mirrors.

The group was intended to be a training or self-analytic group whose task was to study its own functioning. My role was simply to facilitate the process, and I provided little structure. I managed to get a coleader, an advanced graduate student, highly recommended by a colleague of mine.

I carefully explained to the class the purpose and intent of this type of group. They assured me they understood and were excited by the experiential opportunity.

The first session was unsettling for the group members. With no clear agenda, they floundered. Their requests for direction were reflected back to them. The anxiety level remained high, inhibiting the group's movement. My coleader was terrified and quickly identified with the group members, abandoning me. From there, it got worse.

The processing sessions were no better. All 42 of us sat in a circle to review the session, and became yet another group. In spite of their doubts, I remained steadfast in my determination to provide this wonderful experience for them. As the group went on, it became painfully clear that we were stuck and they were furious with me. They felt helpless and exposed in the fishbowl, and I was doing nothing to help them. My coleader was also angry with me, although she and I spent hours processing the sessions.

No matter what I tried, I was unable to get them to acknowledge their frustration and anger with the lack of structure and lack of leadership. The group was stuck and so was I.

One member was very disruptive during these sessions. She continually questioned the group's motives; she arrived 15 minutes late for every session; she was clearly angry at me and in a state of total denial. She refused to express any feelings or take responsibility for her behavior. In hindsight, it was evident that she was frightened, but at the time I was unable to see it. I tried in vain to have the group address this issue. One session, she was absent from the group and failed to let us know she would not be there. This was a rule the group had agreed on. When she returned the next session, I expressed my anger at her behavior and told her it was disruptive to the group. It was evident she was furious with me, but when asked for her feelings she said she was "fine." The group mobilized to her defense. Only two members felt that her behavior, her tardiness, and her absenteeism were disruptive to the group, and they remained very quiet.

During the process session, observers felt I had been unfair toward her and too critical of her. I was angry at her behavior, and that anger combined with my own ineffectiveness may have immobilized the group. As a result, the group never recovered, and politely finished their sessions.

I was distraught, and had, on a regular basis, been calling my former supervisor to discuss the weekly sessions. I tried to process the experience with the class as best I could, examining all aspects of the group and my behavior. Although the students were polite, I clearly turned some of them off to the group experience, the very thing I had been so critical of during my own graduate training. I gave a lot of thought to that class, and I still remember it clearly. In reviewing what happened, I came to several conclusions.

First, the class was too large. It was impossible to attend to all of the feelings of the observers that had been generated by the group. Second, my presence in the fishbowl inhibited students' freedom of expression. They were intimidated by me and unable to reveal their anger, fearing, I suspect, some retribution from me. Third, my coleader was poorly trained and frightened by the unstructured group experience. Finally, I had difficulty moving from the fishbowl to the classroom. The emotional impact of the group was considerable, and it was with great difficulty that I suppressed those feelings during the classroom discussion.

I completely restructured the next course. I limited the number of students to 18, and I refrained from leading any more fishbowl groups. I decided to teach an Advanced Group Class conjointly with the Introduction to Group course. I selected eight students who were interested in leading groups. Throughout the year, those students would facilitate the fishbowl groups.

During the fall quarter, I ran three groups, each comprised of six student members and two coleaders from the advanced group class. Each week, one group would meet in the fishbowl during the first hour of class. This group was observed by me and the remainder of the class. The other two groups met following the class period. Then the following week another group would meet in the fishbowl, so that each group was observed three times over the quarter. This format gave me an opportunity to point out various group processes during the sessions, and it seemed to me that all students participated equally in the course. The advanced group students met later in the evening for their supervisory sessions. It was exhausting work, but a definite improvement over my first model.

However, there were still difficulties. My presence was still an inhibiting factor for students. Even though I was no longer in the fishbowl, my proximity behind the glass had a dampening effect on their actions. Furthermore, students were inhibited by the observations of their classmates, because they were unable to discover how they were perceived after having self-disclosed in the fishbowl. There was no opportunity for processing feelings between those in the fishbowl and those who observed. The feedback sessions were guarded, with each group quick to defend itself during processing. So the search for an effective method of teaching group dynamics continued.

One winter quarter, I tried a leaderless group. There were 10 students in the class. They worked in the fishbowl, and I observed through the glass. We processed each session immediately following the group, but it went nowhere. The members remained inhibited and very little learning took place. Finally, after several years of trial and error, I arrived at a method for providing an effective group experience.

The class is divided into three parts: lecture, group, and processing. The groups meet each week for 10 weeks, and are led by two coleaders from the Advanced Group Class. However, they are not observed. Following each group session, students are given time to write in a journal about the group experience. They are asked to record their feelings about the experience, along with their observations of the group's dynamics. They are encouraged to integrate their assigned reading material into their writing. The journals are turned in to me each week. During class, we discuss reading assignments, lecture material, and observations about group dynamics drawn from their weekly groups. The discussion of group process avoids any recounting of the group's "story." The discussions are much freer than in previous models, because students are now the experts teaching me about their group's dynamics. The supervisory sessions for the advanced group students are held later in the evening.

After processing the class, reading the journals, and hearing the advanced students' accounts of the sessions, I am able to piece together the development process in each group. My lecture for the following week is then constructed in anticipation of the dynamics that each group will face in the coming week. I try to predict what information will be germane to the group's level of development. Teaching group dynamics, using this model, has been successful for me. It provides students with a monitored group experience while providing them a safe arena for exploration of personal issues.

Recently I was invited to teach a group course at another university. Previously, the course had been taught in two sections, an experiential component in which students "role played" group members and the more formal classroom part where group theory and research was discussed. Role playing can offer students a safe means of protecting themselves from self-revelation. Under the facade of the role played, the student might feel freer to participate in a group tied to a graduate course. However, it is also very difficult and even confusing in emotionally charged situations to keep clear boundaries between the role played and the true self, especially when the group experience and role are recurring week to week. In my experience, role-playing can last for about 4 minutes and then students revert to playing themselves. It seems more honest and less confusing to allow them to do so from the group's inception. Instructors sometimes use this model to protect their ethical boundaries in teaching a group course in which the student is just role-playing. However, it seems to me that any competent instructor can quickly see through the role-play to the student's real self. So what has really been gained by this artificial teaching method is a kind of liability insurance for the instructor. Better to drop the games and be honest with students and allow them to interact freely and directly without any instructor involvement.

When I taught the course, I created a schedule in which leadership was rotated among the members on a weekly basis. In addition, I assigned a commentator who was directed to lead the processing of the group for the last 15 minutes of their scheduled hour. Prior to the group experience, the entire class met for 5 minutes to talk about one or two stage-appropriate group issues. Following the groups, we all met for 15 minutes to answer any questions that had to do with group process. The students turned in weekly group papers describing their observations about group development that I commented on and returned the following week. Each week the new leaders were instructed to design a brief introductory exercise that dovetailed with the group's level of maturity. Overall these five groups, each with six or seven members, were highly successful. The students were very motivated and, in part, felt some initial safety because they all had some prior relationship with one another.

It was the best learning experience any students had in one of my group courses. In the more formal class that followed, their understanding of theory and process was clarified and enhanced by their group experiences. Without exception, student feedback revealed this experience to be positive. In large part it was due to the competency of these students. They were bright and curious, and many had strong senses of their own personal boundaries. With a different mix of students, the class may have had a different outcome. All this is to say that even with the best model, success is not guaranteed. There are many factors that influence the outcome that are beyond the instructor's control. It is best to be open, honest, and direct about the positive and negative possibilities of a group experience before it begins. Groups should be safe and monitored to insure that students are protected. Beyond that, honest feedback and a debriefing should follow the experience.

Teaching experiential courses such as groups requires an emotional investment on the part of the instructor. Each must carefully weigh the positive and negative consequences of direct involvement with students in this kind of learning activity. The obligation to insure proper knowledge and understanding of the power of groups must be balanced against the students' right to privacy. There is no one correct method. Recent ethical guidelines establish strict criteria for dual relationships and should be considered before constructing group training experiences.

This brings me to the second part of this discussion. Should students, as part of their curriculum, be required to take part in a group experience? Absolutely. However, the student's privacy must be protected, and the experience designed for maximum safety. Most master's level programs offer only one course in group counseling. An experiential component should be required. The experience of being a client develops a healthy respect for the power inherent in the leader's role.

CONCLUSION

The ideas that compose chaos theory are not new. They have existed for centuries. However, with the advent of supercomputers, we are seeing them for the first time. Ideas that for many for us have always felt intuitively correct are now being confirmed by science. Although chaos theory has yet to solve anything, it does shine a new light on the inner workings of the universe and the relationship among all things. Most importantly, it gives credibility to the necessity for disorder. Finally, chaos theory, from an ecological perspective, once again confirms for us that although competition plays a vital role in evolution, it is cooperation that will ultimately ensure our survival.

Using chaos theory to interpret small group development provides an innovative perspective from which to consider how groups change and evolve. It underscores the importance of conflict and conflict resolution in developing groups, and highlights the inherent capacity of groups to self-organize if properly contained. Simultaneously, chaos theory helps us recognize that authoritarian and controlling leadership, in any form, severely curtails the creative potential of groups.

The other subjects in this book, from metaphors to transpersonal development, are included to emphasize and broaden awareness of the multifaceted and multilayered character of groups. When taken together, the ideas in this book are intended to invigorate further thought and discussion about the possibilities and potentials of working in small groups.

Notes

Chapter 1

1. Chamberlain, L. (1995). Strange attractors in family patterns. In R. Robertson & A. Combs (Eds.), *Chaos theory in psychology and the life sciences* (pp. 267–273). Mahwah, NJ: Lawrence Erlbaum Associates.
2. Ibid.
3. Butz, M. R. (1992b). Chaos, an omen or transcendence in the psychotherapeutic process. *Psychological Reports, 71*, 827–843.
4. Loye, D., & Eisler, R. (1987). Chaos and transformation: Implications of nonequilibrium theory for social science and society. *Behavioral Science, 32*, 53–65.
5. Robertson, R. (1995). Chaos theory and the relationship between psychology and science. In R. Robertson & A. Combs (Eds.), *Chaos theory in psychology and the life sciences* (pp. 3–15). Mahwah, NJ: Lawrence Erlbaum Associates.
6. Loye & Eisler, 1987.
7. Ibid.
8. Ibid.
9. Lorenz, E. (1964). The problem of deducing the climate from the governing equations. *Tellus, 16*, 1–11.
10. Hooper, J., & Teresi, D. (1986). *The three-pound universe.* New York: Macmillan.
11. Gottman, J. M. (1991). Chaos and regulated change in families: A metaphor for the study of transitions. In P. A. Cowan & M. Hetherington (Eds.), *Family transitions* (pp. 247–272). Hillsdale, NJ: Lawrence Erlbaum Associates.
12. Hooper & Teresi, 1986.
13. Rico, G. L. (1991). *Pain and possibility.* New York: Jeremy Tarcher.
14. Barton, S. (1994). Chaos, self-organization, and psychology. *American Psychologist, 49*(1), 5–14.
15. Goerner, S. (1995). Chaos, evolution, and deep ecology. In R. Robertson & A. Combs (Eds.), *Chaos theory in psychology and the life sciences* (pp. 17–38). Mahwah, NJ: Lawrence Erlbaum Associates.

16. Barton, 1994.
17. Goerner, 1995, p. 24.
18. Marion, R. (1992). Chaos, typology, and social organization. *Journal of School Leadership*, 2, 149.
19. Goerner, 1995, p. 19.
20. Barton, 1994.
21. Marion, 1992.
22. Barton, 1994.
    Abraham, F. D., Abraham, R. H., & Shaw, C. D. (1990). *A visual introduction to dynamical systems theory for psychology*. Santa Cruz, CA: Aerial.
    Tufillaro, N. B., Abbott, T., & Reilly, J. (1992). *An experimental approach to nonlinear dynamics and chaos*. Reading, MA: Addison Wesley.
23. Vann Spruiell, M. (1993). Deterministic chaos and the sciences of complexity: Psychoanalysis in the midst of a general scientific revolution. *Journal of the American Psychoanalytic Association, 41*(1), 3–44.
24. Ibid.
25. Marion, 1992.
26. Ibid.
27. Hooper & Teresi, 1986.
28. Ruelle, D., & Takens, F. (1971). On the nature of turbulence. *Communications in Mathematical Physics, 20*, 167–192.
29. Hooper & Teresi, 1986.
30. Ibid.
31. Fuhriman, A., & Burlingame, G. M. (1994). Measuring small group process: A methodological application of chaos theory. [Special Issue: Research problems and methodology]. *Small Group Research, 25*(4), 502–519.
    Burlingame, G. M., Fuhriman, A., & Barnum, K. R. (1995). Group therapy as a nonlinear dynamic system: Analysis of therapeutic communication for chaotic patterns. In F. D. Abraham & A. R. Gilgen (Eds.), *Chaos theory in psychology*. Westport, CT: Praeger Publishers/Greenwood Publishing Group.
32. Gleick, J. (1987). *Chaos: Making a new science*. New York: Viking Penguin.
33. Ibid., p. 8.
34. Hooper & Teresi, 1986.
35. Young, T. R. (1995). Chaos theory and social dynamics: Foundations of postmodern social science. In R. Robertson & A. Combs (Eds.), *Chaos theory in psychology and the life sciences* (pp. 217–233). Mahwah, NJ: Lawrence Erlbaum Associates.
36. Ibid.
37. Jantsch, E. (1980). *The self-organizing universe*. New York: Pergamon.
    Jantsch, E. (1982). *The evolutionary vision*. Boulder: Westview.
    Kauffman, S. A. (1993). *The origins of order*. New York: Oxford University Press.
    Prigogine, I., & Stengers, I., (1984). *Order out of chaos*. New York: Bantam Books.
38. Barton, 1994, p. 7.
39. Jantsch, 1980.
    Jantsch, 1982.
40. Prigogine & Stengers, 1984
41. Hooper & Teresi, 1986.
42. Goerner, 1995, p. 24.
43. Hooper & Teresi, 1984.
44. Ibid.
45. Glance, N., & Huberman, B. (1994, March). The dynamics of social dilemmas. *Scientific American, 270*(3), 76–81.
46. Goerner, 1995, p. 26.

47. Lovelock, J. (1979). *Gaia: A new look at life on earth.* Oxford, England: Oxford University Press.
48. Barton, 1994.
49. Schore, N. E. (1981). Chemistry and human awareness: Natural scientific connections. In R. S. Valle & R. von Eckartsburg (Eds.), *The metaphors of consciousness* (pp. 437–460). New York: Plenum.
50. Barton, 1994.
    Hooper & Teresi, 1986.
51. Schore, 1981.
52. Loye & Eisler, 1987.
53. Jantsch, 1980.
54. Corning, P. (1983). *The synergism hypothesis.* New York: McGraw-Hill.
55. Allen, T. (1986). Newsletters of the SGSR-SIG on hierarchy theory. *General Systems Bulletin, 16*(2), 54–57.
    Voorhees, B. (1986). Toward duality theory. *General Systems Bulletin, 16*(2), 58–62.
56. Allen, 1986.
57. Henderson, H. (1981). *The politics of the solar age.* New York: Doubleday/Anchor.
58. Loye & Eisler, 1987.
59. Ibid.
60. Ibid., p. 58.

Chapter 2

1. Prigogine & Stengers, 1984.
2. Ibid., p. 52.
3. Loye & Eisler, 1987.
4. Durkheim, E. (1964). *The rules of sociological method.* New York: Free Press.
5. Durkheim, E. (1951). *Suicide.* New York: Free Press.
6. Loye, D. (1977).
7. Sorokin, P. (1941). *The crisis of our age.* New York: Dutton.
8. Riegel, K. (1979). *Foundations of dialectical psychology.* New York: Academic.
9. Spengler, O. (1932). *The decline of the west.* New York: Knopf.
   Toynbee, A. (1947). *A study of history.* New York: Oxford University Press.
10. Abraham, F. D., & Gilgen, A. R. (Eds.). (1995). *Chaos theory in psychology.* Westport, CT: Praeger Publishers/Greenwood Publishing Group.
    Duke, M. P. (1994). Chaos theory and psychology: Seven propositions. *Genetic, Social, and General Psychology Monographs, 120*(3), 267–286.
    Kincanon, E., & Powel, W. (1995). Chaotic analysis in psychology and psychoanalysis. *Journal of Psychology, 129*(5), 495–505.
    Krippner, S. (1994). Humanistic psychology and chaos theory: The third revolution and the third force. Society for Chaos Theory in Psychology Inaugural Conference. *Journal of Humanistic Psychology, 34*(3), 48–61.
11. Hager, D. (1992). Chaos and growth. *Psychotherapy, 29,* 378–384.
    Lonie, I. (1991). Chaos theory: A new paradigm for psychotherapy? *Australian and New Zealand Journal of Psychiatry, 25*(4), 548–560.
    Meehl, P. (1978). Theoretical risks and tabular asterisks: Sir Karl, Sir Ronald, and the slow process of soft psychology. *Journal of Consulting and Clinical Psychology, 46,* 806–834.
    Wieland-Burston, J. (1992). *Chaos and order in the world of the psyche.* London: Routledge.
12. Berchulski, S., Conforti, M., Guiter-Mazer, I., & Malone, J. (1995). Chaotic attractors in the therapeutic system. *World Futures, 44*(2-3), 101–114.
    Chamberlain, L. (1990). Chaos and the butterfly effect in family systems. *Network, 8*(3), 11–12.

Elkaim, M. (1990). *If you love me, don't love me: Construction of reality and change in family therapy.* New York: Basic Books.

Kaplan, D. (Ed.). (1995). Systems theory across the Counseling Program [Special issue]. *Journal for the Professional Counselor, 10*(2).

Minuchin, S., & Fishman, H. C. (1981). *Family therapy techniques.* Cambridge, MA: Harvard University Press.

Stevens, B. A. (1991). Chaos: A challenge to refine systems theory. *Australian and New Zealand Journal of Family Therapy, 12*(1), 23–26.

Ward, M. (1995). Butterflies and bifurcations: Can chaos theory contribute to our understanding of family systems? *Journal of Marriage and Family, 57*(3), 629–638.

13. Freeman, W. (1990). Searching for signal and noise in the chaos of the brain waves. In S. Krasner (Ed.), *The uniquity of chaos* (pp. 47–55). Washington, DC: American Association for the Advancement of Science.

Freeman, W. (1991, February). The physiology of perception. *Scientific American, 264,* 78–85.

Vandervert, L. R. (1995). Chaos theory and the evolution of consciousness and mind: A thermodynamic-halographic resolution to the mind–body problem. *New Ideas in Psychology, 13*(2), 107–127.

14. Galatzer-Levy, R. M. (1995). Psychoanalysis and dynamical systems theory: Prediction and self similarity. *Journal of the American Psychoanalytic Association, 43*(4), 1085–1113.

Grotstein, J. S. (1990). Nothingness, meaninglessness, chaos, and the "black hole": Vol. 1. The importance of nothingness, meaninglessness and chaos in psychoanalysis. *Contemporary Psychoanalysis, 26,* 257–290.

Langs, R. (1992). Towards building psychoanalytically based mathematical models or psychotherapeutic paradigms. In R. L. Levine & H. E. Fitzgerald (Eds.), *Analysis of dynamic psychological systems* (Vol. 2, pp. 371–393). New York: Plenum.

Levinson, E. A. (1994). The uses of disorder: Chaos theory and psychoanalysis. *Contemporary Psychoanalysis, 30*(1), 5–24.

Moran, M. G. (1991). Chaos theory and psychoanalysis: The fluidic nature of the mind. *International Review of Psychoanalysis, 18,* 211–221.

15. Putnam, F. (1988). The switch process in multiple-personality disorder and other state-change disorders. *Dissociation, 1,* 24–32.

Putnam, F. (1989). *Diagnosis and treatment of multiple personality disorder.* New York: Guilford.

16. Paulus, M. P., Geyer, M. A., & Braff, D. L. (1996). Use of methods from chaos theory to quantify a fundamental dysfunction in the behavioral organization of schizophrenic patients. *American Journal of Psychiatry, 153*(5), 714–717.

Schmid, G. B. (1991). Chaos, theory, and schizophrenia: Elementary aspects. *Psychopathology, 24,* 185–198.

17. Freedman, A. M.(1995). The biopsychosocial paradigm and the future of psychiatry. *Comprehensive Psychiatry, 36*(6), 397–406.

Sabelli, H. C., & Carlson-Sabelli, L. (1989). Biological priority and psychological supremacy: A new integrative paradigm derived from process theory. *American Journal of Psychiatry, 146,* 1541–1551.

18. Abraham, Abraham, & Shaw (1990).

Butz, M. R. (1992a). The fractal nature of the development of the self. *Psychological Reports, 71,* 1043–1063.

Butz, 1992b.

Van Eenwyk, J. R. (1991). Archetypes: The strange attractors of the psyche. *Journal of Analytic Psychology, 36,* 1–25.

19. Glover, H. (1992). Emotional numbing: A possible endorphin-mediated phenomenon associated with post-traumatic stress disorders and other allied psychopathological states.

Journal of Traumatic Stress, 5, 643–675.

Gurian, P. H., Elliot, E., & Everett, D. (1996). An application of nonlinear dynamics to the presidential nomination process. *Behavioral Science, 41*(4), 271–290.

20. Iannone, R. (1995). Chaos theory and its implications for curriculum and teaching. *Education, 115*(4), 541–558.

21. Brack, C. J., Brack, G., & Zucker, A. (1995). How chaos and complexity theory can help counselors to be more effective. *Counseling and Values, 39*(3), 200–208.

Brennan, C. (1995). Beyond theory and practice: A postmodern perspective. [Special Issue: Rethinking uncertainty and chaos: Possibilities for counseling]. *Counseling and Values, 39*(2), 99–107.

Gelatt, H. B. (1995). Chaos and compassion. [Special Issue: Rethinking uncertainty and chaos: Possibilities for counseling]. *Counseling and Values, 39*(2), 108–116.

Miller, M. J. (1995). A case for uncertainty in career counseling. *Counseling and Values, 39*(3), 162–168.

Wilbur, M. P., Kulikowich, J. M., Roberts-Wilbur, J., & Torres-Rivera, E. (1995). Chaos theory and counselor training. [Special Issue: Rethinking uncertainty and chaos: Possibilities for counseling]. *Counseling and Values, 39*(2), 129–144.

22. Crossan, M. M., Lane, H. W., White, R. E., & Klus, L. (1996). The improvising organization: Where planning meets opportunity. *Organizational Dynamics, 24*(4), 20–35.

23. Gregersen, H. B., & Sailer, L. (1993). Chaos theory and its implications for social science research. *Human Relations, 46*(7), 777–802.

24. von Bertalanffy, L. (1968). *General systems theory.* New York: George Braziller.

25. Lewin, K. (1951). *Field theory in social science* (pp. 228–229). New York: Harper Row.

26. Goldstein, J. (1995). The tower of Babel in nonlinear dynamics: Toward the clarification of terms. In R. Robertson & A. Combs (Eds.), *Chaos theory in psychology and the life sciences* (pp. 39–47). Mahwah, NJ: Lawrence Erlbaum Associates.

27. Marion, 1992.

28. Freeman, 1990.

29. Barton, 1994.

30. Ibid., p. 10.

31. Ibid., p. 12.

32. Freeman, 1990.

33. Freeman, 1991.

34. Barton, 1994.

35. Ibid.

36. Ibid., p. 8.

37. Ibid.

38. Ibid., p. 9.

39. Ibid.

40. Abraham, Abraham, & Shaw, 1990.
Elkaim, 1990.

41. Prigogine, I. (1977).

42. Laszlo, E. (1987). *Evolution: The grand synthesis.* Boston: Shambhala.

43. Elkaim, 1990.

44. Ibid., p. 120.

45. Ibid., p. 120.

46. Butz, 1992a.
Butz, 1992b.
Van Eenwyk, 1991.

47. Butz, 1992a.
Butz, 1992b.

48. Freud, S. (1949). *An outline of psycho-analysis.* New York: Norton.

49. Erickson, E. (1950). *Childhood and society.* New York: Norton.
    Erickson, E. (1982). *The life cycle completed.* New York: Norton.
50. Jung, C. G. (1969). *The structure and dynamics of the psyche.* In *Collected works of C. G. Jung* (Vol. 6). Princeton, NJ: Princeton University Press.
51. Hager, D. (1992). Chaos and growth. *Psychotherapy, 29,* 378–384.
52. Erickson, 1950.
    Erickson, 1982.
53. Erickson, 1950.
    Erickson, 1982.
54. Jung, C. G. (1969). Structure and dynamics of the psyche. In C. G. Jung, G. Adler, & R. F. Hull (Eds. & Trans.), *Collected works of C. G. Jung* (Vol. 8, p. 608). Princeton, NJ: Princeton University Press.
55. Freud, 1949.
56. Jung, C. G. (1971). Psychological types. In C. J. Jung, G. Adler, & R. F. Hull (Eds. & Trans.), *Collected works of C. G. Jung* (Vol. 6, p. 632). Princeton, NJ: Princeton University Press.
57. Freeman, 1990.
    Freeman, 1991.
58. Prigogine & Stengers, 1984.
59. Elderidge, N., & Gould, S. (1972). Punctuated equilibria: An alternative to phyletic gradualism. In T. J. Schopf (Ed.), *Models in paleobiology.* San Francisco: Freeman, Cooper.
60. Maturana, H., & Varela, F. (1980). *Autopoiesis and cognition: The realization of the living.* Boston: Reidel.
61. Jung, 1971.
    Jung, C. G. (1970). Mysterium coniunctionis. In C. G. Jung, G. Adler, & R. F. Hull (Eds. & Trans.), *Collected works of C. G. Jung* (Vol. 14, p. 438). Princeton, NJ: Princeton University Press.
    Loye & Eisler, 1987, p. 56.
62. Jung, 1971.
63. Butz, 1992a, pp. 1060–1061.
64. Winnicott, D. W. (1965). *Maturational process and the facilitating environment.* London: Hogarth.
65. Safan-Gerard, D. (1985). Chaos and control in the creative process. *Journal of the American Academy of Psychoanalysis, 13*(1), 133.
66. Gottman, 1991.
67. Watzlawick, P., Weakland, J., & Fisch, R. (1974). *Change.* New York: Norton.
68. Butz, 1992a.
    Butz, 1992b.
    Gemmill, G., & Wynkoop, C. (1991). The psychodynamics of small group transformation. *Small Group Research, 22,* 4–23.
69. Gottman, 1991, pp. 249–250.
70. Ibid., p. 262.
71. Ibid., p. 262.
72. Ibid.
73. Ibid.
74. Conrad, M. (1986). What is the use of chaos? In A. V. Holden (Ed.), *Chaos.* Princeton, NJ: Princeton University Press.
    Kauffman, S. A. (1991). Antichaos and adaptation. *Scientific American, 265,* 78–84, 118.
75. Butz, 1992a.
    Conrad, 1986.
    Wheatley, M. (1992). *Leadership and the new science.* San Francisco: Berrett-Koehler.

76. Wheatley, 1992.
77. Laszlo, 1987, p. 98.
78. Ibid., p. 98.

Chapter 3

1. Rogers, C. (1970). *Carl Rogers on encounter groups*. New York: Harper & Row.
2. Ibid., p. 113.
3. Bennis, W. G. (1964). Patterns and vicissitudes in T-group development. In L. Bradford, J. Gibb, & K. Benne (Eds.), *T-group theory and laboratory method: Innovation in re-education*. New York: Wiley.
4. Rogers, 1970.
5. Cissna, K. (1984). Phases in group development: The negative evidence. *Small Group Behavior, 15*(1), 3–32.
6. McClure, B. A. (1994). The shadow side of regressive groups. *Counseling and Values, 38,* 76–88.

   McClure, B. A. (1990). The group mind: Generative and regressive groups. *Journal for Specialists in Group Work, 15*(3), 159–170.
7. Kaplan, S., & Roman, M. (1963). Phases of development in adult therapy group. *International Journal of Group Psychotherapy, 13,* 10–26.

   Mills, T. (1964). *Group transformation: An analysis of a learning group*. Englewood Cliffs, NJ: Prentice-Hall.

   Tuckman, B. (1965). Developmental sequence in small groups. *Psychological Bulletin, 63,* 384–399.

   Tuckman, B., & Jensen, M. (1977). Stages of small group development revisited. *Group and Organizational Studies, 2*(4), 419–427.

   Yalom, I. D. (1970). *The theory and practice of group psychotherapy*. New York: Basic Books.

   Braaten, L. (1975). Developmental phases of encounter groups and related intensive groups. *Interpersonal Development, 5,* 112–129.

   Caple, R. (1978). The sequential stages of group development. *Small Group Behavior, 9*(4), 470–476.

   Gatz, Y., & Christie, L. (1991). Marital group metaphors: Significance in the life stages of group development. *Contemporary Family Therapy: An International Journal, 12*(1), 23–26.

   Levine, B. (1979). *Group psychotherapy*. New York: Waveland Press.

   Sarri, R., & Galinsky, M. (1985). A conceptual framework for group development. In M. Sundel (Ed.), *Individual change through small group* (2nd ed.). New York: Free Press.

   Napier, R. W., & Gershenfeld, M. K. (1985). *Groups* (3rd ed.). Boston: Houghton Mifflin.

   Browner, A. (1989). Group development as constructed social reality: A social-cognitive understanding of group formation. *Social Work With Groups, 12*(2), 23–41.

   Posthuma, B. (1989). *Small groups in therapy settings: Process and leadership*. Boston: College-Hill.

   Trotzer, J. (1989). *The counselor and the group*. New York: Free Press.

   Ettin, M. (1992). *Foundations and applications of group psychotherapy*. Boston: Allyn & Bacon.

   MacKenzie, K., & Livesley, J. (1990). *Introduction to time-limited group psychotherapy*. Washington, DC: American Psychiatric Press.

   Wheelan, S. (1990). *Facilitating training groups*. New York: Praeger.

   Rutan, J., & Stone, W. (1993). *Psychodynamic group psychotherapy* (2nd ed.). New York: Guilford.
8. Ettin, 1992.

9. Kaplan & Roman, 1963; Tuckman, 1965; Yalom, 1970.
10. Tuckman, 1965; Caple, 1978; Sarri & Galinsky, 1985; Napier, 1987; Posthuma, 1989; MacKenzie & Livesley, 1990; Wheelan, 1990.
11. Mills, 1964; Braaten, 1975; Tuckman & Jensen, 1977; Caple, 1978; Levine, 1979; Sarri & Galinsky, 1985; MacKenzie & Livesley, 1990; Trotzer, 1989; Wheelan, 1990; Ettin, 1992; Rutan & Stone, 1993.
12. McClure, 1990, 1994.
13. Young, A. (1976). *The reflexive universe.* Lake Oswego, OR: Robert Briggs Associates.
14. Yalom, 1970.
15. MacKenzie & Livesley, 1990.
16. Wheelan, 1990.
17. Posthuma, 1989.
18. Napier & Geishenfeld, 1987.
19. Mills, 1964.
20. McClure, 1990, 1994.
21. Wheelan, 1990.
22. Yalom, 1970; Braaten, 1975; Trotzer, 1989.
23. Wheelan, 1990; Posthuma, 1989; Rutan & Stone, 1993.
24. Braaten, 1975; Sarri & Galinsky, 1985; Ettin, 1992.
25. Lonie, I. (1995). The princess and the swineherd: Applications of chaos theory to psychodynamics. In R. Robertson & A. Combs (Eds.), *Chaos theory in psychology and the life sciences* (pp. 285–294). Mahwah, NJ: Lawrence Erlbaum Associates.
26. Hansen, J. C., Warner, R. W., & Smith, E. J. (1980). *Group counseling: Theory and practice* (2nd ed., p. 513). Boston: Houghton Mifflin.
27. Shantz, C. U. (1987). Conflicts between children. *Child Development, 58,* 283–305.
28. Yalom, I. D. (1985). *The theory and practice of group psychotherapy* (3rd ed., p. 352). New York: Basic Books.
29. Cowger, C. D. (1979). Conflict and conflict management in working with groups. *Social Work with Groups, 2*(4), 309–320.
30. Jandt, F. E. (1976). *The process of interpersonal communications.* New York: Harper & Row.
31. McClure, B. A. (1989). More metaphor: considerations and concerns in groups. *Small Group Behavior, 20*(4), 449–458.
32. Peck, M. S. (1983). *People of the lie* (p. 32). New York: Simon & Schuster.
33. Briggs, J., & Peat, F. D. (1989). *Turbulent mirror: An illustrated guide to choas theory and the science of wholeness* (p. 165). New York: Harper & Row.
34. Shlain, L. (1991). *Art and physics* (p. 34). New York: Quill.
35. Ibid.
36. Cohen, L. (1993). *Stranger music* (p. 373). New York: Vintage Books.
37. Rico, P. (1991). *Pain and possibility.* New York: Jeremy Tarcher.
38. Ibid.
39. Ibid., p. 32.
40. Ibid.
41. Briggs & Peat, p. 194.
42. Ibid.

Chapter 4

1. Briggs, J., & Peat, F. D. (1989). *Turbulent mirror: An illustrated guide to chaos theory and the science of wholeness.* New York: Harper & Row.
2. Hooper, J., & Teresi, D. (1986). *The three-pound universe.* New York: Macmillan.
3. Ibid.

4. Briggs & Peat, 1989.
5. Senge, P. (1990). *The fifth discipline*. New York: Doubleday.
6. Briggs & Peat, 1989.
7. Ibid.
8. Ibid., p. 194.
9. Ibid., p. 193.
10. Gemmill, G., & Wynkoop, C. (1991). The psychodynamics of small group transformation. *Small Group Research, 22*, 4–23.
11. Young, T. R. (1995). Chaos theory and social dynamics: Foundations of postmodern social science. In R. Robertson & A. Combs (Eds.), *Chaos theory in psychology and the life sciences* (pp. 217–233). Mahwah, NJ: Lawrence Erlbaum Associates.
12. Ibid., p. 228.
13. Ibid., p. 229.
14. Gemmill & Wynkoop, 1991.
15. Ibid., p. 6.
16. Ibid., p. 13.
17. Shlain, L. (1991). *Art and physics*. New York: Quill.
18. Ibid.

Chapter 5

1. Humphrey, L., & Stern, S. (1988). Object relations and the family system in bulimia: A theoretical integration. *Journal of Marital and Family Therapy, 14*(4), 337–350.
2. Bion, W. R. (1961). *Experiences in groups and other papers*. New York: Basic Books.
3. Vann Spruiell, M. (1993). Deterministic chaos and the sciences of complexity: Psychoanalysis in the midst of a general scientific revolution. *Journal of the American Psychoanalytic Association, 41*(1), 3–44.
4. Senge, P. (1990). *The fifth discipline* (p. 242). New York: Doubleday.
5. Goertzel, B. (1995). Belief system as attractors. In R. Robertson & A. Combs (Eds.), *Chaos theory in psychology and the life sciences* (pp. 123–134). Mahwah, NJ: Lawrence Erlbaum Associates.
6. Chamberlain, L. (1995). Strange attractors in patterns of family interaction. In R. Robertson & A. Combs (Eds.), *Chaos theory in psychology and the life sciences* (pp. 267–273). Mahwah, NJ: Lawrence Erlbaum Associates.
7. Cambel, A. B. (1992). *Applied chaos theory: A paradigm for complexity*. San Diego: Academic.
8. Senge, 1990.
9. Yalom, I. D. (1985). *The theory and practice of group psychotherapy* (3rd ed.). New York: Basic Books.
10. Johnson-Laird, P. (1988). Freedom and constrain in creativity. In R. J. Steinberg (Ed.), *The nature of creativity*. New York: Cambridge University Press.
11. Ibid.
12. Bayes, M., & Newton, P. (1978). Women in authority: A sociopsychological analysis. *The Journal of Applied Behavioral Science, 14*(1), 7–20.
Bernardez, T. (1982). The female therapist in relation to male roles. In K. Solomon & N. B. Levy (Eds.), *Men in transition: Theories and therapies for psychological health*. New York: Plenum.
Bernardez, T. (1983). Women in authority: Psychodynamic and interactional aspects. In B. G. Reed & C. D. Gavin (Eds.) (1983), *Group work with women/groupwork with men: An overview of gender issues in social groupwork practice*. New York: Hawthorne Press.
Bernardez, T. (1996). Conflicts with anger and power in women's Groups. In B. DeChant (Ed.), *Women and group psychotherapy: Theory and practice* (pp. 171–199). New York: Guilford.

DeChant, B. (Ed.), *Women and group psychotherapy: Theory and practice* (pp. 171–199). New York: Guilford.

Elrick, M. (1977). The leader, she: Dynamics of a female-led self-analytic group. *Human Relations, 30*(10), 869–878.

Hagen, B. H. (1983) Managing conflict in all-women groups. In B. G. Reed & C. D. Gavin (Eds.), *Group work with women/groupwork with men: an overview of gender issues in social groupwork practice*. New York: Hawthorne.

Hulse, D. (1985). Overcoming the social-ecological barriers to group effectiveness: present and future. *Journal for Specialists in Group Work, 10*(2), 92–97.

Martin, P. Y., & Shanahan, K. A. (1983). Transcending the effects of sex composition in small groups. In B. G. Reed & C. D. Gavin (Eds.), *Group work with women/groupwork with men: An overview of gender issues in social groupwork practice*. New York: Hawthorne.

McWilliams, N., & Stein, J. (1987). Women's groups led by women: The management of devaluing transferences. *The International Journal of Group Psychotherapy, 37*(2), 139–153.

Reed, B. G. (1983). Women leaders in small groups: Social-psychological perspectives and strategies. In B. G. Reed & C. D. Gavin (Eds.), *Group work with women/groupwork with men: An overview of gender issues in social groupwork practice*. New York: Hawthorne.

Gavin, C. D., & Reed, B. G. (1983). Gender issues in social group work: An overview. In B. G. Reed & C. D. Gavin (Eds.), *Group work with women/groupwork with men: an overview of gender issues in social groupwork practice*. New York: Hawthorne.

13. Alexander, D., & Andersen, K. (1993). Gender as a factor in the attribution of leadership traits. *Political Science Quarterly, 46*, 527–545.

Bayes & Newton, 1978.

Elrich, 1977.

Page, B. (1996). Gender issues and leadership: How to improve communication among and between men and women. In *The Olympics of Leadership: Overcoming Obstacles, Balancing Skills, Taking Risks. Proceedings of the Annual International Conference of the National Community College Chair Academy* (Phoenix, AZ, February 14–17).

Schoenholtz-Read, J. (1996). Sex-role issues: Mixed-gender therapy groups as the treatment of choice. In B. DeChant (Ed.), *Women and group psychotherapy: Theory and practice* (pp. 223–241). New York: Guilford.

Silver, A. L. (1996). Women who lead. *American Journal of Psychoanalysis, 56*(1), 3–16.

14. Bayes & Newton, 1978.

Bernardz, T. (1996). Gender-based countertransference in the group treatment of women. In B. DeChant (Ed.), *Women and group psychotherapy: Theory and practice* (pp. 400–424). New York: Guilford.

Cohn, B. R. (1996). Narcissism in women in groups: The emerging female self. In B. DeChant (Ed.), *Women and group psychotherapy: Theory and practice* (pp. 157–175). New York: Guilford.

Rosenberg, P. (1996). Comparative leadership styles of male and female therapists. In B. DeChant (Ed.), *Women and group psychotherapy: Theory and practice* (pp. 425–441). New York: Guilford.

15. Bayes & Newton, 1978.

Kahn, E. W. (1996). Coleadership gender issues in group psychotherapy. In B. DeChant (Ed.), *Women and group psychotherapy: Theory and practice* (pp. 442–461). New York: Guilford.

16. Chodorow, N., & Contratto, S. (1982). The fantasy of the perfect mother. In B. Thorne (Ed.), *Rethinking the family* (pp. 54–75). New York: Longman.

17. Reed, 1983.

18. Offermann, L. R., & Beil, C. (1992). Achievement styles of women leaders and their peers: Toward understanding of women and leadership. *Psychology of Women Quarterly, 16*(1), 37–56.

19. Reed, 1983, p. 37.
20. Ibid., p. 38.
21. Reed, 1983.
22. Hagen, 1983.
23. Ibid.
24. McWilliams & Stein, 1987.
25. Ibid., pp. 142–143.
26. McWillaims & Stein, 1987.
27. Bayes & Newton, 1978, p. 19.
28. Bernardez, 1983.
    Bernardez, T. (1996). Conflicts with anger and power in women's Groups. In B. DeChant (Ed.), *Women and group psychotherapy: Theory and practice* (pp. 171–199). New York: Guilford.
29. Hagen, 1983.
30. Ibid.
31. Bernardez, 1983.
32. Ibid.
    Schoenhotz-Read, J. (1996). Sex-role issues: Mixed-gender therapy groups as the treatment of choice. In B. DeChant (Ed.), *Women and group psychotherapy: Theory and practice* (pp. 223–241). New York: Guilford.
33. Bernardez, 1983, p. 46.
34. Elrick, 1977.
35. Bayes & Newton, 1978, p. 14.

Chapter 6

1. Singer, E. (1970). *Key concepts in psychotherapy* (2nd ed.). New York: Basic Books.

Chapter 8

1. Davis, J. M., & Sandoval, J. (1978). Metaphor in group health consultation. *Journal of Community Psychology, 6,* 374–382.
2. Bion, W. R. (1961). *Experiences in groups and other papers.* New York: Basic Books.
3. Bales, R. F. (1970). *Personality and interpersonal behavior.* New York: Holt, Rinehart & Winston.
   Slater, P. E. (1966). *Microcosm: Structural, psychological and religious evolution in groups.* New York: Wiley.
4. Cobb, C. M. (1973). *Group metaphor: Its function, form, theme, and evolution.* Unpublished doctoral dissertation, Harvard Graduate School of Education.
   Davis, J. M., & Sandoval, J. (1978). Metaphor in group health consultation. *Journal of Community Psychology, 6,* 374–382.
   Ettin, M. F. (1986). Within the group's view: Clarifying dynamics through metaphoric and symbolic imagery. *Small Group Behavior, 17*(4), 407–426.
   Gladding, S. T. (1984). The metaphor as a counseling tool in group work. *Journal for Specialists in Group Work, 9*(3), 151–156.
   McClure, B. A. (1987). Metaphoric illumination in groups. *Small Group Behavior, 18*(2), 179–187.
   Owen, W. F. (1985). Metaphor analysis of cohesiveness in small discussion groups. *Small Group Behavior, 16*(3), 415–424.
5. Davis & Sandoval, 1978, p. 374.
6. Rossel, R. D. (1981). Word play: Metaphor and humor in the small group. *Small Group Behavior, 12*(1), 116–136.

7. Davis & Sandoval, 1978.
   Rossel, 1981.
8. Davis & Sandoval, 1978.
9. Ibid.
   McClure, 1987.
10. Ettin, 1986.
    McClure, 1987.
11. McClure, 1987.
    Rossel, 1981.
12. Ettin, 1986.
13. Ettin, 1986.
    McClure, 1987.
    Morocco, C. C. (1979). The development and function of group metaphor. *Journal of Theory of Social Behaviour, 9*, 15–27.
    Rossel, 1981.
14. Jung, C. G. (1971). Psychological types. In C. G. Jung, G. Adler, & R. F. Hull (Eds. & Trans.), *Collected works of C. G. Jung* (Vol. 6, p. 632). Princeton, NJ: Princeton University Press.
15. Morocco, 1979.
16. Ibid., p. 16.
17. Hansen, J., Warner, R., & Smith, E. (1976). *Group counseling: Theory and process.* Chicago: Rand-McNally.
18. Rossel, 1981, p. 122.
19. Ibid.
20. Davis & Sandoval, 1978.
21. Ettin, 1986, p. 422.
22. Katz, G. A. (1983). The noninterpretation of metaphors in psychiatric hospital groups. *International Journal of Group Psychotherapy, 33*(1), 53–67.
23. Davis & Sandoval, 1978.
24. McClure, 1987.
25. Rossel, 1981, p. 119.
26. Ettin, 1986; Katz, 1983; Rossel, 1981.
27. Katz, 1983, p. 61.

Chapter 9

1. Lebon, G. (1908). *The crowd: A study of the popular mind* (pp. 14–16). London: T. Fisher Unwin.
2. O'Hara, M. M., & Wood, J. K. (1983). Patterns of awareness: consciousness and the group mind. *The Gestalt Journal, 6*(2), 103–116.
3. Horwitz, L. (1983). Projective identification in dyads and groups. *International Journal of Group Psychotherapy, 33*(3), 277.
4. McClure, B. (1990). The group mind: Generative and regressive groups. *Journal for Specialists in Group Work, 15*(3), 159–170.
5. Elmes, M., & Gemmill, G. (1990). The psychodynamics of mindlessness and dissent in small groups. *Small Group Research, 21*(1), 28–44.
6. Goerner, S. (1995). Chaos, evolution, and deep ecology. In R. Robertson & A. Combs (Eds.), *Chaos theory in psychology and the life sciences* (pp. 17–38). Mahwah, NJ: Lawrence Erlbaum Associates.
7. McClure, 1990.
8. Gemmill, G. (1986). The dynamics of the group shadow in intergroup relations. *Small Group Behavior, 17*(2), 229–240.

9. McClure, 1990.

10. Kormanski, C. (1988). Using group development theory in business and industry. *Journal for Specialists in Group Work, 3*(1), 30–43.

11. Yalom, I. D. (1985). *The theory and practice of group psychotherapy* (3rd ed.). New York: Basic Books.

12. McPhail, C. (1991). *The myth of the madding crowd.* New York: Aldine De Gruyter.

13. Lebon, 1908.
    Freud, S. (1955). *Group psychology and the analysis of ego.* New York: Liverright Publishing Corporation.
    Lewin, K. (1959). *Field theory in social science.* London: Tavistock.

14. Lebon, 1908.

15. McClure, B., Miller, T., & Russo, T. (1992). Conflict within a children's group: Suggestions for facilitating its expression and resolution strategies. *The School Counselor, 39*(4), 268–272.

16. Staub, E. (1989). *The roots of evil.* New York: Cambridge University Press.

17. Deikman, A. (1991). *The wrong way home. Uncovering cult behavior in American society.* Boston, MA: Beacon.
    McClure, 1990.

18. Grosso, M. (1985). *The final choice.* Walpole, NH: Stillpoint.
    Menninger, K. (1938). *Man against himself.* New York: Harcourt, Brace.

19. McClure, 1990.

20. Peck, M. S. (1983). *People of the lie.* New York: Simon & Schuster.

21. Ibid., p. 223.

22. Deikman, 1991.

23. Ibid.

24. Peck, 1983.

25. Ibid.
    Deikman, 1991.

26. Elmes, M., & Gemmill, G. (1990). The psychodynamics of mindlessness and dissent in small groups. *Small Group Research, 21*(1), 28–44.

27. Gemmill, 1986.
    Peck, 1983.
    Staub, 1989.

28. Peck, 1983.

29. Gemmill, 1986.

30. Peck, 1983.
    Staub, 1989.

31. Gemmill, 1986.
    Staub, 1989.
    Peck, 1983.

32. Elmes & Gemmill, 1990.

33. Deikman, 1991.

34. Peck, 1985, p. 220.

35. Golding, W. G. (1954). *The lord of the flies.* New York: Coward-McCann.

36. McClure, 1990.

37. Deikman, 1991.

38. Johnson, H. (1991). *Sleep walking through history: America throughout the Reagan years.* New York: Norton.

39. Woodward, B. (1991). *The commanders.* New York: Simon & Schuster.

40. Ad Hoc Media Group, 1991.

41. MacArthur, J. (1992). *Second front.* New York: Hill & Wang.
   Gottschalk, M. (1992, Summer). Operation desert cloud: The media and the gulf war. *World Policy Journal, 9*(3), 449.
42. Solomon, N. (1991, May 24). The media protest too much. *New York Times*, p. 31.
43. Gottschalk, 1992.
44. Ibid.
45. MacArthur, 1992, p. 45.
46. Staub, 1989.
47. Greenpeace, 1992.
   Gottschalk, 1992.
48. Gelman, 1991.
49. Postol, 1992.
50. Draper, 1992.
51. Greenpeace. (1992). *The body count.* Washington, DC: Author.
52. Lewis, 1992.
53. Royce, 1991.
54. Atkins, E. (1992, April 17). Campus racism unrest spreads. *Detroit News*, p. 1.
   Salter, C. (1992, January 24). Hatred against Jews on the rise, Anti-Defamation League warns. *Atlanta Constitution*, p. 9.
55. Staub, 1989.
56. Elmes & Gemmill, 1990.
57. McClure, 1990.
58. McClure, B. A. (1989a). More metaphor: Concerns and consideration in groups. *Small Group Behavior, 20*(4), 449–458.
   McClure, B. A. (1989b). What's a group metaphor? *Journal for Specialists in Group Work, 14*, 239–242.
   McClure, B. A. (1987). Metaphoric illumination in groups. *Small Group Behavior, 18*(2), 179–187.
   Barrett, F., & Cooperrider, D. (1990). Generative metaphor intervention: A new approach for working with systems divided by conflict and caught in defensive perception. *Journal of Applied Behavioral Science, 26*(2), 219–239.
59. Elmes & Gemmill, 1990.
60. Fynsk, in Scott, J. (1992). The campaign against political correctness: What's really at stake. *Radical History Review, 54*, 59–80.
61. Scott, 1992, p. 77.
62. Scott, 1992, p. 77.
63. Raikka, J. (1997). On dissociating oneself from collective responsibility. *Social Theory and Practice, 23*(1), 93–99.
64. Nemeth, C. (1985). Dissent, group process and creativity: The contribution of minority influence. In E. Lawler (Ed.), *Advances in group processes* (pp. 57–75). Greenwich, CT: JAI.

Chapter 10

1. Bion, W. R. (1961). *Experiences in groups and other papers.* New York: Basic Books.
   Cattell, R. J. (1938). *Psychology and the religious quest.* New York: Nelson.
   Grosso, M. (1985). *The final choice.* Walpole, NH: Stillpoint.
   Hartstone, C. (1942). Elements of truth in the group-mind concept. *Social Research, 9*, 248–259.
   McDougall, 1920.

2. Freud, S. (1955). *Group psychology and the analysis of ego.* New York: Liverright.
   Lebon, G. (1908). *The crowd: A study of the popular mind.* London: T. Fisher Unwin.
   Lewin, K. (1959). *Field theory in social science.* London: Tavistock.
3. Conforti, M. (1997). The generation of order and form within transpersonal fields: Insights from the psychotherapeutic situation. [Special issue: The concept of the collective consciousness: Research perspectives]. *World Futures, 48*(1–4), 171–191.
   Sagi, M. (1997). Holistic healing as fresh evidence for collective consciousness. [Special issue: The concept of the collective consciousness: Research perspectives]. *World Futures, 48*(1–4), 151–161.
   Schlitz, M. J. (1997). Intentionality: An argument for transpersonal consciousness distant. [Special issue: The concept of the collective consciousness: Research perspectives]. *World Futures, 48*(1–4), 115–127.
   Varvoglis, M. P. (1997). Conceptual frameworks for the study of transpersonal consciousness. [Special issue: The concept of the collective consciousness: Research perspectives]. *World Futures, 48*(1–4), 105–124.
4. Bion, 1961.
   Cattell, 1938.
   Durkheim, E. (1951). *Suicide.* New York: Free Press.
   Grof, S. (1976). *Realms of the human unconscious.* New York: Dutton.
   Grosso, 1975.
   Hartstone, C. (1942). Elements of truth in the group-mind concept. *Social Research, 9,* 248–259.
   Horwitz, L. (1983). Projective identification in dyads and groups. *International Journal of Group Psychotherapy, 33*(3), 259–279.
   Huxley, A. (1954). *Doors of perception.* New York: Harper & Brothers.
   Jantsch, E. (1980). *The self-organizing universe.* New York: Pergamon.
   Lewin, 1959.
   McDougall, 1920.
   Randell, J. (1975). *Parapsychology and the nature of life.* New York: Harper & Row.
   Rubinstein, D. (1982). Individual minds and the social order. *Qualitative Sociology, 5*(2), 121–139.
   Welwood, J. (1979). Self-knowledge as the basis for an integrative psychology. *The Journal of Transpersonal Psychology, 11*(1), 23–40.
5. Post, D. L. (1980). Floyd H. Allport and the launching of modern social psychology. *Journal of the History of the Behavioral Sciences, 16,* 369–376.
6. Bion, 1961.
7. Hobbs, 1951.
8. Davis, J. M., & Sandoval, J. (1978). Metaphor in group health consultation. *Journal of Community Psychology, 6,* 374–382.
9. Conforti, 1997.
   Sagi, 1997.
   Schlitz, 1997.
   Varvoglis, 1997.
10. Capra, F. (1988). *Uncommon wisdom.* New York: Simon & Schuster.
    Grosso, 1985.
    Welwood, 1979.
11. Kaplan, H. I., & Sadock, B. J. (Eds.). (1972). *Sensitivity through encounter and marathon* (p. xx). New York: Dutton.
12. Catlin & Parsons, cited in Rubenstein, 1982, pp. 121–122.
13. McDougall, 1920; Bion, 1960.
14. Bion, 1961, p. 65.
15. McDougall, 1920.

16. Post, 1980.
17. Allport, cited in Post, 1980.
18. Allport, p. 371.
19. Post, 1980, p. 372.
20. Kaplan & Sadock, 1972, p. xx.
21. Hartstone, 1942.
22. Conforti, 1997.
    Grof, 1976.
    Grosso, 1975.
    O'Hara, M. M., & Wood, J. K. (1983). Patterns of awareness: consciousness and the group mind. *The Gestalt Journal,* 6(2), 103–116.
    Sagi, 1997.
    Schlitz, 1997.
    Varvoglis, 1997.
    Welwood, 1979.
23. Grosso, 1985, p. 200.
24. Ibid.
25. Grof, cited in Capra, 1988.
    O'Hara & Wood, 1983.
26. Grosso, 1985, p. 201.
27. Ibid., p. 218.
28. Hartstone, 1954.
29. Randall, 1975, p. 235.
30. Lovelock, J. (1979). *Gaia: A new look at life on earth.* Oxford, England: Oxford University Press.
31. Grof, 1976.
    Welwood, 1979.
32. McDougall, 1920.
33. Koestler, A. (1972). *The roots of coincidence* (p. 108). New York: Random House.
34. Ibid., p. 130.
35. Huxley, 1954, p. 22.
36. Ibid., 1954.
37. Cattell, 1938, p. 64.
38. Jantsch, 1975.
39. Grosso, 1985.
40. Loye, D. (1983). *The sphinx and the rainbow.* Boulder, CO: Shambhala.
41. Ibid., p. 125.
42. Ibid., p. 126.
43. Teilhard de Chardin, P. (1959). *The phenomenon of man.* New York: Harper & Row.
44. Combs, A., & Holland, M. (1990). *Synchronicity.* New York: Paragon House.
45. Ibid., p. 42.
46. Ibid., p. 42.
47. Grosso, 1985.
48. Progoff, I. (1973). *Jung, synchronicity, and human destiny.* New York: Dell.
49. Loye, 1983.
50. Ibid., 1983.
51. Jung, in Loye, 1983, p. 126.
52. O'Hara & Wood, 1983, p. 113.
53. O'Hara & Wood, 1983.
54. Ibid., p. 109.
55. Maslow, A. (1971). *The farther reaches of human nature.* New York: Viking Press.

56. Jung, C. G. (1971). Psychological types. In C. G. Jung, G. Adler, & R. F. Hull (Eds. & Trans.), *Collected works of C. G. Jung* (Vol. 6, p. 632). Princeton, NJ: Princeton University Press.
57. Progoff, I. (1953). *Jung's psychology and its social meaning.* London: Routledge and Kegan Paul.
58. Progroff, 1973.
59. Combs & Holland, 1990. p. 73
60. Grosso, 1985, p. 201.
61. Alexander, E. (1991). Personal communication.
62. Combs & Holland, 1990, p. 74.
63. Ibid., p. 31.
64. Jung, C. G. (1969). Structure and dynamics of the psyche. In C. G. Jung, G. Adler, & R. F. Hull (Eds. & Trans.), *Collected works of C. G. Jung* (Vol. 8, p. 608). Princeton, NJ: Princeton University Press.
65. Shelburne, W. (1988). *Mythos and logos in the thought of Carl Jung* (p. 65). Albany: State University of New York Press.
66. Combs & Holland, 1990.
67. O'Hara & Wood, 1983, p. 108.
68. Metzner, R. (1980). Ten classical metaphors of self-transformation. *The Journal of Transpersonal Psychology, 12*(1), 47–62.
69. Bales, R. F. (1970). *Personality and interpersonal behavior.* New York: Holt, Rinehart & Winston.
70. Ibid., p. 151.
71. Campbell, 1988, p. 22.
72. Ibid., p. 29.
73. Ibid., p. 32.
74. Combs & Holland, 1990, p. xxx.
75. Padfield, S. (1980). Mind–matter interaction in the psychokinetic experience. In B. D. Josephson & V. S. Ramachandran (Eds.), *Consciousness and the physical world.* New York: Pergamon.
76. Loye, 1983.
77. Bohm, D. (1980). *Wholeness and the implicate order.* London: Routledge & Kegan Paul.
78. Loye, 1983, p. 150.
79. Bohm, 1980, p. 145.
80. Loye, 1983, p. 144.
81. Ibid., p. 151.
82. Ibid., p. 154.
83. Ibid.
84. Ibid.
85. Ibid.
86. Loye, 1983.
87. Combs & Holland, 1990.
88. Cited in Combs & Holland, 1990, p. 45.
89. Loye, 1983.
90. Ibid., p. 159.
91. Ibid.
92. Padfield, S. (1981). Archetypes: Synchronicity and manifestation. *Psychoenergetics, 4,* 77–81.
    Padfield, 1980.
93. Combs & Holland, 1990.
94. Loye, 1983, p. 153.
95. Huxley, A. (1954). *Doors of perception* (p. 153). New York: Harper & Brothers.

96. Sheldrake, R. (1991, February/March). The rebirth of nature. *Creation and Spirituality, 7,* 16–17.
97. McClure, 1990.
98. Watzlawick, Beavin, & Jackson, 1967.
99. O'Hara & Wood, 1983.
100. Grosso, 1985.
101. Phillips, 1986.
102. Halling, S., Rowe, J. O., Davies, E., Leifer, M., Powers, D., & van Bronkhorst, J. (1986, May). *The psychology of forgiving another* (p. 14). Paper presented at the Western Psychological Association, Seattle.
103. Napier, A., & Whitaker, C. (1978). *The family crucible.* New York: Harper & Row.
104. Safan-Gerard, D. (1985). Chaos and control in the creative process. *Journal of the American Academy of Psychoanalysis, 13*(1), 131.
105. Ibid., p. 135.
106. O'Hara & Wood, 1983.
107. Maslow, 1972, p. 271.
108. O'Hara & Wood, 1983, p. 108.
109. Perls, Hefferline, & Goodman, cited in O'Hara & Wood, 1983.
110. O'Hara & Wood, 1983.
111. Owen, I. M., & Sparrow, M. (1976). *Conjuring up Phillip.* New York: Pocket Books.
112. Ibid.
113. Ibid.

Chapter 11

1. Slater, P. E. (1966). *Microcosm: Structural, psychological and religious evolution in groups* (p. 12). New York: Wiley.
2. Trotzer, J. (1989). *The counselor and the group.* New York: Free Press.

References

Abraham, F. D., Abraham, R. H., & Shaw, C. D. (1990). *A visual introduction to dynamical systems theory for psychology.* Santa Cruz, CA: Aerial.

Abraham, F. D., & Gilgen, A. R. (Eds.). (1995). *Chaos theory in psychology.* Westport, CT: Praeger Publishers/Greenwood Publishing Group.

Adams, H. A., & Chadbourne, J. (1982, April). Therapeutic metaphor: An approach to weight control. *The Personnel and Guidance Journal, 8*(60), 510–512.

Ad Hoc Media Group. (1991, June 25). Letter to Dick Cheney, Secretary of Defense.

Alexander, D., & Andersen, K. (1993). Gender as a factor in the attribution of leadership traits. *Political Science Quarterly, 46,* 527–545.

Allen, T. (1986). Newsletters of the SGSR-SIG on hierarchy theory. *General Systems Bulletin, 16*(2), 54–57.

Atkins, E. (1992, April 17). Campus racism unrest spreads. *Detroit News,* p. 1.

Bales, R. F. (1970). *Personality and interpersonal behavior.* New York: Holt, Rinehart & Winston.

Barrett, F., & Cooperrider, D. (1990). Generative metaphor intervention: A new approach for working with systems divided by conflict and caught in defensive perception. *Journal of Applied Behavioral Science, 26*(2), 219–239.

Barton, S. (1994). Chaos, self-organization, and psychology. *American Psychologist, 49*(1), 5–14.

Bateson, M. C. (1989). *Composing a life.* New York: Penguin.

Bayes, M., & Newton, P. (1978). Women in authority: A sociopsychological analysis. *The Journal of Applied Behavioral Science, 14*(1), 7–20.

Bennis, W. G. (1964). Patterns and vicissitudes in T-group development. In L. Bradford, J. Gibb, & K. Benne (Eds.), *T-group theory and laboratory method: Innovation in re-education.* New York: Wiley.

Berchulski, S., Conforti, M., Guiter-Mazer, I., & Malone, J. (1995). Chaotic attractors in the therapeutic system. *World Futures, 44*(2–3), 101–114.

Bernardez, T. (1982). The female therapist in relation to male roles. In K. Solomon & N. B. Levy (Eds.), *Men in transition: Theories and therapies for psychological health.* New York: Plenum.

Bernardez, T. (1983). Women in authority: Psychodynamic and interactional aspects. In B. G. Reed & C. D. Gavin (Eds.), *Group work with women/group work with men: An overview of gender issues in social group work practice.* New York: Hawthorne.

231

Bernardez, T. (1996). Conflicts with anger and power in women's Groups. In B. DeChant (Ed.), *Women and group psychotherapy: Theory and practice* (pp. 171–199). New York: Guilford.

Bernardez, T. (1996). Gender-based countertransference in the group treatment of women. In B. DeChant (Ed.), *Women and group psychotherapy: Theory and practice* (pp. 400–424). New York: Guilford.

Billow, R. M. (1981). Observing spontaneous metaphor in children. *Journal of Experimental Child Psychology, 31*, 430–445.

Bion, W. R. (1961). *Experiences in groups and other papers.* New York: Basic Books.

Bohm, D. (1980). *Wholeness and the implicate order.* London: Routledge & Kegan Paul.

Bonner, H. (1959). *Group dynamics: Principles and applications.* New York: Ronald Press.

Braaten, L. (1975). Developmental phases of encounter groups and related intensive groups. *Interpersonal Development, 5*, 112–129.

Brack, C. J., Brack, G., & Zucker, A. (1995). How chaos and complexity theory can help counselors to be more effective. *Counseling and Values, 39*(3), 200–208.

Brennan, C. (1995). Beyond theory and practice: A postmodern perspective. [Special Issue: Rethinking uncertainty and chaos: Possibilities for counseling]. *Counseling and Values, 39*(2), 99–107.

Briggs, J., & Peat, F. D. (1989). *Turbulent mirror: An illustrated guide to chaos theory and the science of wholeness.* New York: Harper & Row.

Browner, A. (1989). Group development as constructed social reality: A social-cognitive understanding of group formation. *Social Work With Groups, 12*(2), 23–41.

Burlingame, G. M., Fuhriman, A., & Barnum, K. R. (1995). Group therapy as a nonlinear dynamic system: Analysis of therapeutic communication for chaotic patterns. In F. D. Abraham & A. R. Gilgen (Eds.), *Chaos theory in psychology.* Westport, CT: Praeger Publishers/Greenwood Publishing Group.

Butz, M. R. (1992a). The fractal nature of the development of the self. *Psychological Reports, 71*, 1043–1063.

Butz, M. R. (1992b). Chaos, an omen or transcendence in the psychotherapeutic process. *Psychological Reports, 71*, 827–843.

Butz, M., Chamberlain, L., & McCown, W. (1997). *Strange attractors: Chaos, complexity, and the art of family therapy.* New York: Wiley.

Cambel, A. B. (1992). *Applied chaos theory: A paradigm for complexity.* San Diego: Academic.

Campbell, J. (1988). *Inner reaches of outer space: Metaphor as myth and religion.* New York: Van der Marck.

Caple, R. (1978). The sequential stages of group development. *Small Group Behavior, 9*(4), 470–476.

Capra, F. (1988). *Uncommon wisdom.* New York: Simon & Schuster.

Cattell, R. J. (1938). *Psychology and the religious quest.* New York: Nelson.

Chamberlain, L. (1990). Chaos and the butterfly effect in family systems. *Network, 8*(3), 11–12.

Chamberlain, L. (1995). Strange attractors in family patterns. In R. Robertson & A. Combs (Eds.), *Chaos theory in psychology and the life sciences* (pp. 267–273). Mahwah, NJ: Lawrence Erlbaum Associates.

Chessick, R. D. (1969). Was Machiavelli right? *American Journal of Psychotherapy, 23*(4), 633–644.

Chodorow, N., & Contratto, S. (1982). The fantasy of the perfect mother. In B. Thorne (Ed.), *Rethinking the family* (pp. 54–75). New York: Longman.

Cissna, K. (1984). Phases in group development: The negative evidence. *Small Group Behavior, 15*(1), 3–32.

Cobb, C. M. (1973). *Group metaphor: Its function, form, theme, and evolution.* Unpublished doctoral dissertation, Harvard University, Graduate School of Education.

Cohen, L. (1993). *Stranger music.* New York: Vintage Books.

Cohn, B. R. (1996). Narcissism in women in groups: The emerging female self. In B. DeChant (Ed.), *Women and group psychotherapy: Theory and practice* (pp. 157–175). New York: Guilford.

Combs, A., & Holland, M. (1990). *Synchronicity.* New York: Paragon.

Conforti, M. (1997). The generation of order and form within transpersonal fields: Insights from the psychotherapeutic situation. [Special issue: The concept of the collective consciousness: Research perspectives]. *World Futures, 48*(1–4), 171–191.

Conrad, M. (1986). What is the use of chaos? In A. V. Holden (Ed.), *Chaos.* Princeton, NJ: Princeton University Press.

Corning, P. (1983). *The synergism hypothesis.* New York: McGraw-Hill.

Cowger, C. D. (1979). Conflict and conflict management in working with groups. *Social Work with Groups, 2*(4), 309–320.

Crossan, M. M., Lane, H. W., White, R. E., & Klus, L. (1996). The improvising organization: Where planning meets opportunity. *Organizational Dynamics, 24*(4), 20–35.

Csanyi, V., & Kampis, G. (1985). Autogensis: Evolution of replicative systems. *Journal of Theoretical Biology, 114*, 303–321.

Csikszentmihalyi, M. (1993). *Flow: The psychology of optimal experience.* New York: HarperCollins.

Davis, J. M., & Sandoval, J. (1978). Metaphor in group health consultation. *Journal of Community Psychology, 6*, 374–382.

de Chardin, P. (1959). *The phenomenon of man.* New York: Harper & Row.

DeChant, B. (Ed.). (1996). *Women and group psychotherapy: Theory and practice.* New York: Guilford.

Deikman, A. (1991). *The wrong way home. Uncovering cult behavior in American society.* Boston: Beacon.

Draper, T. (1992, January 30). The true history of the Gulf War. *New York Review of Books,* p. 42.

Dunphry, D. (1968). Phases, roles, and myths in self analytic groups. *Journal of Applied Behavioral Science, 4*, 195–225.

Durkheim, E. (1951). *Suicide.* New York: Free Press.

Durkheim, E. (1965). *The rules of sociological method.* New York: Free Press.

Eenwyk, J. R. (1991). Archetypes: The strange attractors of the psyche. *Journal of Analytic Psychology, 36*, 1–25.

Elderidge, N., & Gould, S. (1972). Punctuated equilibria: An alternative to phyletic gradualism. In T. J. Schopf (Ed.), *Models in paleabiology.* San Francisco: Freeman, Cooper.

Elkaim, M. (1990). *If you love me, don't love me: Construction of reality and change in family therapy.* New York: Basic Books.

Elmes, M., & Gemmill, G. (1990). The psychodynamics of mindlessness and dissent in small groups. *Small Group Research, 21*(1), 28–44.

Elrick, M. (1977). The leader, she: Dynamics of a female-led self-analytic group. *Human Relations, 30*(10), 869–878.

Erickson, E. (1950). *Childhood and society.* New York: Norton.

Erickson, E. (1982). *The life cycle completed.* New York: Norton.

Ettin, M. (1992). *Foundations and applications of group psychotherapy.* Boston: Allyn & Bacon.

Ettin, M. F. (1986). Within the group's view: Clarifying dynamics through metaphoric and symbolic imagery. *Small Group Behavior, 17*(4), 407–426.

Freedman, A. M. (1995). The biopsychosocial paradigm and the future of psychiatry. *Comprehensive Psychiatry, 36*(6), 397–406.

Freeman, W. (1990). Searching for signal and noise in the chaos of the brain waves. In S. Krasner (Ed.), *The uniquity of chaos* (pp. 47–55). Washington, DC: American Association for the Advancement of Science.

Freeman, W. (1991, February). The physiology of perception. *Scientific American, 264*, 78–85.

Freud, S. (1949). *An outline of psycho-analysis.* New York: Norton.

Freud, S. (1955). *Group psychology and the analysis of ego.* New York: Liverright.

Fuhriman, A., & Burlingame, G. M. (1994). Measuring small group process: A methodological application of chaos theory. [Special Issue: Research problems and methodology]. *Small Group Research, 25*(4), 502–519.

Galatzer-Levy, R. M. (1995). Psychoanalysis and dynamical systems theory: Prediction and self similarity. *Journal of the American Psychoanalytic Association, 43*(4), 1085–1113.

Garfinkel, A. (1983). A mathematics for physiology. *American Journal of Physiology, 245,* 455–466.

Garvin, C. D., & Reed, B. G. (1983). Gender issues in social group work: An overview. In B. G. Reed & C. D. Gavin (Eds.), *Group work with women/group work with men.* New York: Hawthorne.

Gelatt, H. B. (1995). Chaos and compassion. [Special Issue: Rethinking uncertainty and chaos: Possibilities for counseling]. *Counseling and Values, 39*(2), 108–116.

Gellman, B. (1991, March 16). U.S. bombs missed 70% of time. *Washington Post,* p. 16.

Gemmill, G. (1986). The dynamics of the group shadow in intergroup relations. *Small Group Behavior, 17*(2), 229–240.

Gemmill, G., & Wynkoop, C. (1991). The psychodynamics of small group transformation. *Small Group Research, 22,* 4–23.

Gregersen, H. B., & Sailer, L. (1993). Chaos theory and its implications for social science research. *Human Relations, 46*(7), 777–802.

Gladding, S. T. (1984). The metaphor as a counseling tool in group work. *Journal for Specialists in Group Work, 9*(3), 151–156.

Glance, N., & Huberman, B. (1994, March). The dynamics of social dilemmas. *Scientific American, 270*(3), 76–81.

Gleick, J. (1987). *Chaos: Making a new science.* New York: Viking Penguin.

Glover, H. (1992). Emotional numbing: A possible endorphin-mediated phenomenon associated with post-traumatic stress disorders and other allied psychopathological states. *Journal of Traumatic Stress, 5,* 643–675.

Goerner, S. (1995). Chaos, evolution, and deep ecology. In R. Robertson & A. Combs (Eds.), *Chaos theory in psychology and the life sciences* (pp. 17–38). Mahwah, NJ: Lawrence Erlbaum Associates.

Goertzel, B. (1995). Belief system as attractors. In R. Robertson & A. Combs (Eds.), *Chaos theory in psychology and the life sciences* (pp. 123–134). Mahwah, NJ: Lawrence Erlbaum Associates.

Golding, W. G. (1954). *The lord of the flies.* New York: Coward-McCann.

Goldstein, J. (1995). The tower of Babel in nonlinear dynamics: Toward the clarification of terms. In R. Robertson & A. Combs (Eds.), *Chaos theory in psychology and the life sciences* (pp. 39–47). Mahwah, NJ: Lawrence Erlbaum Associates.

Gottman, J. M. (1991). Chaos and regulated change in families: A metaphor for the study of transitions. In P. A. Cowan & M. Hetherington (Eds.), *Family transitions* (pp. 247–272). Hillsdale, NJ: Lawrence Erlbaum Associates.

Gottschalk, M. (1992, Summer). Operation desert cloud: The media and the gulf war. *World Policy Journal, 9*(3), 449.

Green, M. (1985). Talk and doubletalk: The development of metacommunication knowledge about oral language. *Research in the Teaching of English, 19*(1), 9–24.

Greenpeace (1992). *The body count.* Washington, DC: Author.

Grof, S. (1976). *Realms of the human unconscious.* New York: Dutton.

Grosso, M. (1985). *The final choice.* Walpole, NH: Stillpoint.

Grotstein, J. S. (1990). Nothingness, meaninglessness, chaos, and the "black hole": Vol. 1. The importance of nothingness, meaninglessness and chaos in psychoanalysis. *Contemporary Psychoanalysis, 26,* 257–290.

Guidano, V. F. (1991). *The self in process.* New York: Guilford.

Gurian, P. H., Elliot, E., & Everett, D. (1996). An application of nonlinear dynamics to the presidential nomination process. *Behavioral Science, 41*(4), 271–290.

Hagen, B. H. (1983). Managing conflict in all-women groups. In B. G. Reed & C. D. Gavin (Eds.), *Group work with women/group work with men.* New York: Hawthorne.

Hager, D. (1992). Chaos and growth. *Psychotherapy, 29,* 378–384.

Halling, S., Rowe, J. O., Davies, E., Leifer, M., Powers, D., & van Bronkhorst, J. (1986, May). *The psychology of forgiving another.* Paper presented at the Western Psychological Association, Seattle.

Hansen, J., Warner, R., & Smith, E. (1976). *Group counseling: Theory and process.* Chicago: Rand McNally.

Hansen, J., Warner, R., & Smith, E. (1980). *Group counseling: Theory and process* (2nd ed.). Chicago: Rand McNally.

Hartstone, C. (1942). Elements of truth in the group-mind concept. *Social Research, 9,* 248–259.

Henderson, H. (1981). *The politics of the solar age.* New York: Doubleday/Anchor.

Hooper, J., & Teresi, D. (1986). *The three-pound universe.* New York: Macmillan.

Horwitz, L. (1983). Projective identification in dyads and groups. *International Journal of Group Psychotherapy, 33*(3), 259–279.

Hulse, D. (1985). Overcoming the social-ecological barriers to group effectiveness: present and future. *Journal for Specialists in Group Work, 10*(2), 92–97.

Humphrey, L., & Stern, S. (1988). Object relations and the family system in bulimia: A theoretical integration. *Journal of Marital and Family Therapy, 14*(4), 337–350.

Huxley, A. (1954). *Doors of perception.* New York: Harper & Brothers.

Iannone, R. (1995). Chaos theory and its implications for curriculum and teaching. *Education, 115*(4), 541–558.

Jandt, F. E. (1976). *The process of interpersonal communications.* New York: Harper & Row.

Janis, I. (1972). *Victims of groupthink: A psychological study of foreign policy decisions and fiascoes.* Boston: Houghton Mifflin.

Jantsch, E. (1982). *The evolutionary vision.* Boulder: Westview.

Jantsch, E. (1980). *The self-organizing universe.* New York: Pergamon.

Johnson, H. (1991). *Sleepwalking through history: America throughout the Reagan years.* New York: W. W. Norton.

Jung, C. G. (1923). *Psychological types* (H. G. Barnes, Trans.). New York: Harcourt Brace.

Jung, C. G. (1968). Archetypes and the collective unconscious. In C. G. Jung, G. Adler, & R. F. Hull (Eds. & Trans.), *Collected works of C. G. Jung* (Vol. 8). Princeton, NJ: Princeton University Press.

Jung, C. G. (1969). Structure and dynamics of the psyche. In C. G. Jung, G. Adler, & R. F. Hull (Eds. & Trans.), *Collected works of C. G. Jung* (Vol. 8). Princeton, NJ: Princeton University Press.

Jung, C. G. (1970). Mysterium coniunctionis. In C. G. Jung, G. Adler, & R. F. Hull (Eds. & Trans.), *Collected works of C. G. Jung* (Vol. 14). Princeton, NJ: Princeton University Press.

Jung, C. G. (1971). Psychological types. In C. G. Jung, G. Adler, & R. F. Hull (Eds. & Trans.), *Collected works of C. G. Jung* (Vol. 6). Princeton, NJ: Princeton University Press.

Kahn, E. W. (1996). Coleadership gender issues in group psychotherapy. In B. DeChant (Ed.), *Women and group psychotherapy: Theory and practice* (pp. 442–461). New York: Guilford.

Kaplan, H. I., & Sadock, B. J. (Eds.). (1972). *Sensitivity through encounter and marathon.* New York: E. P. Dutton.

Kaplan, S., & Roman, M. (1963). Phases of development in adult therapy group. *International Journal of Group Psychotherapy, 13,* 10–26.

Katz, G. A. (1983). The noninterpretation of metaphors in psychiatric hospital groups. *International Journal of Group Psychotherapy, 33*(1), 53–67.

Kauffman, S. A. (1991). Antichaos and adaptation. *Scientific American, 265,* 78–84.

Kauffman, S. A. (1993). *The origins of order.* New York: Oxford University.

Koestler, A. (1972). *The roots of coincidence.* New York: Random House.

Koestler, A. (1990). *The act of creation.* New York: Penguin.

Kormanski, C. (1988). Using group development theory in business and industry. *Journal for Specialists in Group Work, 3*(1), 30–43.

Langs, R. (1992). Towards building psychoanalytically based mathematical models or psychotherapeutic paradigms. In R. L. Levine & H. E. Fitzgerald (Eds.), *Analysis of dynamic psychological systems* (Vol. 2, pp. 371–393). New York: Plenum.

Laszlo, E. (1987). *Evolution: The grand synthesis.* Boston: Shambhala.

Lebon, G. (1908). *The crowd: A study of the popular mind.* London: T. Fisher Unwin.

Levine, B. (1979). *Group psychotherapy.* New York: Waveland.

Levinson, E. A. (1994). The uses of disorder: Chaos theory and psychoanalysis. *Contemporary Psychoanalysis, 30*(1), 5–24.

Lewin, K. (1951). *Field theory in social science.* New York: Harper Row.

Lewin, K. (1959). *Field theory in social science.* London: Tavistock.

Lonie, I. (1995). The princess and the swineherd: Applications of chaos theory to psychodynamics. In R. Robertson & A. Combs (Eds.), *Chaos theory in psychology and the life sciences* (pp. 285–294). Mahwah, NJ: Lawrence Erlbaum Associates.

Lorenz, E. (1964). The problem of deducing the climate from the governing equations. *Tellus, 16,* 1–11.

Lovelock, J. (1979). *Gaia: A new look at life on earth.* Oxford: Oxford University Press.

Loye, D., & Eisler, R. (1987). Chaos and transformation: Implications of nonequilibrium theory for social science and society. *Behavioral Science, 32,* 53–65.

Loye, D. (1977). *The leadership passion: A psychology of ideology.* San Francisco: Jossey-Bass.

Loye, D. (1983). *The sphinx and the rainbow.* Boulder, Co: Shambhala.

MacArthur, J. (1992). *Second front.* New York: Hill & Wang.

MacKenzie, K., & Livesley, J. (1990). *Introduction to time-limited group psychotherapy.* Washington, DC: American Psychiatric Press.

Maples, M. (1988). Group development: Extending Tuckman's theory. *Journal for Specialists in Group Work, 13,* 17–23.

Marion, R. (1992). Chaos, typology, and social organization. *Journal of School Leadership, 2,* 144–177.

Martin, P. Y., & Shanahan, K. A. (1983). Transcending the effects of sex composition in small groups. In B. G. Reed & C. D. Gavin (Eds.), *Group work with women/group work with men.* New York: Hawthorne.

Maslow, A. (1971). *The farther reaches of human nature.* New York: Viking.

Masterson, J. F. (1988). *The search for the real self: Unmasking the personality disorders of our age.* New York: Free Press.

Maturana, H., & Varela, F. (1980). *Autopoiesis and cognition: The realization of the living.* Boston: Reidel.

McClure, B. (1990). The group mind: Generative and regressive groups. *Journal for Specialists in Group Work, 15*(3), 159–170.

McClure, B., Miller, T., & Russo, T. (1992). Conflict within a children's group: Suggestions for facilitating its expression and resolution strategies, *The School Counselor, 39*(4), 268–272.

McClure, B. A. (1987). Metaphoric illumination in groups. *Small Group Behavior, 18*(2), 179–187.

McClure, B. A. (1989a). More metaphor: Concerns and consideration in groups. *Small Group Behavior, 20*(4), 449–458.

McClure, B. A. (1989b). What's a group metaphor? *Journal for Specialists in Group Work, 14,* 239–242.

McClure, B. A. (1994). The shadow side of regressive groups. *Counseling and Values, 38,* 76–88.

McDougall, W. (1920). *The group mind.* Cambridge, England: Cambridge University Press.

McPhail, C. (1991). *The myth of the madding crowd.* New York: Aldine De Gruyter.

McWilliams, N., & Stein, J. (1987). Women's groups led by women: The management of devaluing transferences. *The International Journal of Group Psychotherapy, 37*(2), 139–153.

Meehl, P. (1978). Theoretical risks and tabular asterisks: Sir Karl, Sir Ronald, and the slow process of soft psychology. *Journal of Consulting and Clinical Psychology, 46*, 806–834.

Menninger, K. (1938). *Man against himself*. New York: Harcourt, Brace.

Metzner, R. (1980). Ten classical metaphors of self-transformation. *The Journal of Transpersonal Psychology, 12*(1), 47–62.

Miller, M. J. (1995). A case for uncertainty in career counseling. *Counseling and Values, 39*(3), 162–168.

Mills, T. (1964). *Group transformation: An analysis of a learning group*. Englewood Cliffs, NJ: Prentice-Hall.

Minuchin, S., & Fishman, H. C. (1981). *Family therapy techniques*. Cambridge, MA: Harvard University Press.

Moran, M. G. (1991). Chaos theory and psychoanalysis: The fluidic nature of the mind. *International Review of Psychoanalysis, 18*, 211–221.

Morocco, C. C. (1979). The development and function of group metaphor. *Journal of Theory of Social Behaviour, 9*, 15–27.

Napier, R. W., & Gershenfeld, M. K. (1985). *Groups* (3rd ed.). Boston: Houghton Mifflin.

Napier, A., & Whitaker, C. (1978). *The family crucible*. New York: Harper & Row.

Nemeth, C. (1985). Dissent, group process and creativity: The contribution of minority influence. In E. Lawler (Ed.), *Advances in group processes* (pp. 57–75). Greenwich, CT: JAI.

Nilsen, D. L. F. (1984). Epiphany: The language of sudden insight. *Exercise Exchange, 30*(1), 13–16.

Offermann, L. R., & Beil, C. (1992). Achievement styles of women leaders and their peers: Toward understanding of women and leadership. *Psychology of Women Quarterly, 16*(1), 37–56.

O'Hara, M. M., & Wood, J. K. (1983). Patterns of awareness: consciousness and the group mind. *The Gestalt Journal, 6*(2), 103–116.

Owen, I. M., & Sparrow, M. (1976). *Conjuring up Phillip*. New York: Pocket Books.

Owen, W. F. (1985). Metaphor analysis of cohesiveness in small discussion groups. *Small Group Behavior, 16*(3), 415–424.

Padfield, S. (1980). Mind–matter interaction in the psychokinetic experience. In B. D. Josephson & V. S. Ramachandran (Eds.), *Consciousness and the physical world*. New York: Pergamon.

Padfield, S. (1981). Archetypes: Synchronicity and manifestation. *Psychoenergetics, 4*, 77–81.

Page, B. (1996). Gender issues and leadership: How to improve communication among and between men and women. In *The Olympics of leadership: Overcoming obstacles, balancing skills, taking risks*. Proceedings of the Annual International Conference of the National Community College Chair Academy, Phoenix.

Paulus, M. P., Geyer, M. A., & Braff, D. L. (1996). Use of methods from chaos theory to quantify a fundamental dysfunction in the behavioral organization of schizophrenic patients. *American Journal of Psychiatry, 153*(5), 714–717.

Peck, M. S. (1983). *People of the lie*. New York: Simon & Schuster.

Peck, M. S. (1987). *The different drum*. New York: Simon & Schuster.

Post, D. L. (1980). Floyd H. Allport and the launching of modern social psychology. *Journal of the History of the Behavioral Sciences, 16*, 369–376.

Posthuma, B. (1989). *Small groups in therapy settings: Process and leadership*. Boston: College-Hill.

Postol, T. (1991–1992 Winter). Lessons of the Gulf War experience with the Patriot. *International Security, 16*(3), 119.

Prigogine, I., & Stengers, I. (1984). *Order out of chaos*. New York: Bantam Books.

Progoff, I. (1953). *Jung's psychology and its social meaning*. London: Routledge and Kegan Paul.

Progoff, I. (1973). *Jung, synchronicity, and human destiny*. New York: Dell.

Purce, J. (1974). *The mystic spiral*. London: Thames & Hudson.

Putnam, F. (1988). The switch process in multiple-personality disorder and other state-change disorders. *Dissociation, 1*, 24–32.

Putnam, F. (1989). *Diagnosis and treatment of multiple personality disorder.* New York: Guilford.

Raikka, J. (1997). On dissociating oneself from collective responsibility. *Social Theory and Practice, 23*(1), 93–99.

Randell, J. (1975). *Parapsychology and the nature of life.* New York: Harper & Row.

Reed, B. G. (1983). Women leaders in small groups: Social–psychological perspectives and strategies. In B. G. Reed & C. D. Gavin (Eds.), *Group work with women/group work with men: An overview of gender issues in social group work practice.* New York: Hawthorne.

Reik, T. (1949). *Masochism in modern man.* New York: Farrar, Straus.

Riegel, K. (1979). *Foundations of dialectical psychology.* New York: Academic.

Rico, G. L. (1991). *Pain and possibility.* New York: Jeremy Tarcher.

Robertson, R. (1995). Chaos theory and the relationship between psychology and science. In R. Robertson & A. Combs (Eds.), *Chaos theory in psychology and the life sciences* (pp. 3–15). Mahwah, NJ: Lawrence Erlbaum Associates.

Robertson, R., & Combs, A. (Eds.). (1995). *Chaos theory in psychology and the life sciences.* Mahwah, NJ: Lawrence Erlbaum Associates.

Rogers, C. (1970). *Carl Rogers on encounter groups.* New York: Harper & Row.

Rosenberg, P. (1996). Comparative leadership styles of male and female therapists. In B. DeChant (Ed.), *Women and group psychotherapy: Theory and practice* (pp. 425–441). New York: Guilford.

Rossel, R. D. (1981). Word play: Metaphor and humor in the small group. *Small Group Behavior, 12*(1), 116–136.

Rubinstein, D. (1982). Individual minds and the social order. *Qualitative Sociology, 5*(2), 121–139.

Ruelle, D., & Takens, F. (1971). On the nature of turbulence. *Communications in Mathematical Physics, 20*, 167–192.

Rutan, J., & Stone, W. (1993). *Psychodynamic group psychotherapy* (2nd ed.). New York: Guilford.

Sabelli, H. C., & Carlson-Sabelli, L. (1989). Biological priority and psychological supremacy: A new integrative paradigm derived from process theory. *American Journal of Psychiatry, 146*, 1541–1551.

Safan-Gerard, D. (1985). Chaos and control in the creative process. *Journal of the American Academy of Psychoanalysis, 13*(1), 129–136.

Sagi, M. (1997). Holistic healing as fresh evidence for collective consciousness. [Special issue: The concept of the collective consciousness: Research perspectives]. *World Futures, 48*(1–4), 151–161.

Salter, C. (1992, January 24). Hatred against Jews on the rise, Anti-Defamation League warns. *Atlanta Constitution*, p. 9.

Sarri, R., & Galinsky, M. (1985). A conceptual framework for group development. In M. Sundel (Ed.), *Individual change through small groups* (2nd ed.). New York: Free Press.

Schlitz, M. J. (1997). Intentionality: An argument for transpersonal consciousness distant. [Special issue: The concept of the collective consciousness: Research perspectives]. *World Futures, 48*(1–4), 115–127.

Schmid, G. B. (1991). Chaos, theory, and schizophrenia: Elementary aspects. *Psychopathology, 24*, 185–198.

Schoenholtz-Read, J. (1996). Sex-role issues: Mixed-gender therapy groups as the treatment of choice. In B. DeChant (Ed.), *Women and group psychotherapy: Theory and practice* (pp. 223–241). New York: Guilford.

Schore, N. E. (1981). Chemistry and human awareness: Natural scientific connections. In R. S. Valle & R. von Eckartsburg (Eds.), *The metaphors of consciousness* (pp. 437–460). New York: Plenum.

Scott, J. (1992). The campaign against political correctness: What's really at stake. *Radical History Review*, 54, 59–80.

Senge, P. (1990). *The fifth discipline*. New York: Doubleday.

Shantz, C. U. (1987). Conflicts between children. *Child Development*, 58, 283–305.

Shelburne, W. (1988). *Mythos and logos in the thought of Carl Jung*. Albany: State University of New York Press.

Sheldrake, R. (1991, February/March). The rebirth of nature. *Creation and Spirituality*, 7, 16–17.

Shlain, L. (1991). *Art and physics*. New York: Quill.

Silver, A. L. (1996). Women who lead. *American Journal of Psychoanalysis*, 56(1), 3–16.

Singer, E. (1970). *Key concepts in psychotherapy* (2nd ed.). New York: Basic Books.

Slater, P. E. (1966). *Microcosm: Structural, psychological and religious evolution in groups*. New York: Wiley.

Solomon, N. (1991, May 24). The media protest too much. *New York Times*, p. 31.

Sorokin, P. (1941). *The crisis of our age*. New York: Dutton.

Spengler, O. (1932). *The decline of the west*. New York: Knopf.

Sprott, J. C. (1993). *Strange attractors: Creating patterns in chaos*. New York: M & T Books.

Stanley-Muchow, J. L. (1985). Metaphoric self-expression in human development. *Journal of Counseling and Development*, 64, 198–201.

Staub, E. (1989). *The roots of evil*. New York: Cambridge University Press.

Stevens, B. A. (1991). Chaos: A challenge to refine systems theory. *Australian and New Zealand Journal of Family Therapy*, 12(1), 23–26.

Toynbee, A. (1947). *A study of history*. New York: Oxford University Press.

Trotzer, J. (1989). *The counselor and the group*. New York: Free Press.

Tuckman, B. (1965). Developmental sequences in small groups. *Psychological Bulletin*, 63(6), 384–399.

Tuckman, B., & Jensen, M. (1977). Stages of small group development revisited. *Group and Organizational Studies*, 2(4), 419–427.

Tufillaro, N. B., Abbott, T., & Reilly, J. (1992). *An experimental approach to nonlinear dynamics and chaos*. Reading, MA: Addison Wesley.

Vandervert, L. (1991). The emergence of brain and mind amid chaos through maximum power evolution. *World Futures: Journal of General Evolution*, 33(4), 253–273.

Vandervert, L. R. (1995). Chaos theory and the evolution of consciousness and mind: A thermodynamic-halographic resolution to the mind-body problem. *New Ideas in Psychology*, 13(2), 107–127.

Van Eenwyk, J. R. (1991). Archetypes: The strange attractors of the psyche. *Journal of Analytical Psychology*, 36, 1–25.

Vann Spruiell, M. (1993). Deterministic chaos and the sciences of complexity: Psychoanalysis in the midst of a general scientific revolution. *Journal of the American Psychoanalytic Association*, 41(1), 3–44.

Varvoglis, M. P. (1997). Conceptual frameworks for the study of transpersonal consciousness. [Special issue: The concept of the collective consciousness: Research perspectives]. *World Futures*, 48(1–4), 105–124.

von Bertalanffy, L. (1969). *General system theory*. New York: George Braziller.

von Bertalanffy, L. (1975). *Perspective on system theory*. New York: George Braziller.

Voorhees, B. (1986). Toward duality theory. *General Systems Bulletin*, 16(2), 58–62.

Ward, M. (1995). Butterflies and bifurcations: Can chaos theory contribute to our understanding of family systems? *Journal of Marriage and Family*, 57(3), 629–638.

Watzlawick, P., Beavin, J. H., & Jackson, D. D. (1967). *Pragmatics of human communication*. New York: Norton.

Watzlawick, P., Weakland, J., & Fisch, R. (1974). *Change*. New York: Norton.

Welch, M. J. (1984). Using metaphor in psychotherapy. *Journal of Psychosocial Nursing, 22*(11), 13–18.

Welwood, J. (1979). Self-knowledge as the basis for an integrative psychology. *The Journal of Transpersonal Psychology, 11*(1), 23–40.

Wheatley, M. (1992). *Leadership and the new science.* San Francisco: Berrett-Koehler.

Wheelan, S. (1990). *Facilitating training groups.* New York: Praeger.

Wilbur, M. P., Kulikowich, J. M., Roberts-Wilbur, J., & Torres-Rivera, E. (1995). Chaos theory and counselor training. [Special Issue: Rethinking uncertainty and chaos: Possibilities for counseling]. *Counseling and Values, 39*(2), 129–144.

Wills, T. A., Weiss, R. L., & Patterson, G. R. (1974). A behavioral analysis of determinants of marital satisfaction. *Journal of Consulting and Clinical Psychology, 42,* 802–811.

Winner, E., Engel, M., & Gardner, H. (1980). Misunderstanding metaphor, what's the problem? *Journal of Experimental Child Psychology, 30,* 20–32.

Winnicott, D. W. (1953). Transitional objects and transitional phenomena. *International Journal of Psycho-analysis, 34,* 89–97.

Winnicott, D. W. (1965). *Maturational Process and the facilitating environment.* London: Hogarth.

Woodward, B. (1991). *The commanders.* New York: Simon & Schuster.

Yalom, I. D. (1970). *The theory and practice of group psychotherapy.* New York: Basic Books.

Yalom, I. D. (1985). *The theory and practice of group psychotherapy* (3rd ed.). New York: Basic Books.

Young, A. (1976). *The reflexive universe.* Lake Oswego, OR: Robert Briggs.

Young, T. R. (1995). Chaos theory and social dynamics: Foundations of postmodern social science. In R. Robertson & A. Combs (Eds.), *Chaos theory in psychology and the life sciences* (pp. 217–233). Mahwah, NJ: Lawrence Erlbaum Associates.

Author Index

241

Subject Index